Essentials for the Scientific and Technical Writer

Hardy Hoover

Editor, Publications Consultant

Instructor, Extension Courses
University of California at Berkeley

Dover Publications, Inc.
New York

Published in Canada by General Publishing Company, Ltd.,
30 Lesmill Road, Don Mills, Toronto, Ontario.
Published in the United Kingdom by Constable and Company,
Ltd., 10 Orange Street, London WC2H 7EG.

This Dover edition, first published in 1980, is a revised and
corrected republication of the work originally published in 1970
by John Wiley & Sons, Inc., under the title *Essentials for the
Technical Writer*.

International Standard Book Number: 0-486-24060-6
Library of Congress Catalog Card Number: 80-66488

Manufactured in the United States of America
Dover Publications, Inc.
180 Varick Street
New York, N.Y. 10014

Preface

This book provides concise, practical help for the scientist, engineer, technician, or student who needs to improve his technical writing. Concentrated explanations and drills are presented which will develop the general writing skill required. Training in a specific science or technology is *not* essential for technical writing, but writing skill in general is a "must."

A broad perspective of the technical writing field is presented so as to alert the reader to the various writing situations he may encounter in his profession. Reports are thoroughly analyzed. And a discussion of the vitally important but little-known specifications ("specs") relates their uses and their value to technical writers in both industry and government.

In order to improve writing techniques at every stage of production (from manuscript planning to publishing), a number of checklists has been included at appropriate places in the text. Beginners, experienced technical writers and editors, and supervisors will all find that these lists improve communication quality and efficiency. They were developed during many years of meeting rigorous publication deadlines.

For either the full-time or the part-time technical writer and editor, the techniques explained and presented in this book will take the guesswork and the rewrite drudgery out of his assignments and will help him to maximize his efficiency. Technical writing is not only a vital part of today's world, but it can also be exciting and rewarding.

Hardy Hoover
April, 1970

Acknowledgments

Of those who aided me in preparing this book, I place first my wife, Ruth, who helped enthusiastically. First-rank adviser was a long-time colleague, William H. Hansen. Others who contributed were Eleanor Schmidt, Clyde C. Coffindaffer, Carol Mayes, D. R. Anderson, R. E. Audette, and Sherol Cantwell. Jay E. Schwartzbine was generosity itself. Charles Alvarez envisaged and helped think out the entire venture.

Contents

The Special Needs
of Technical Writers

The person who pursues a scientific or technical course has a responsibility of which he may not be fully aware: he should be able to write well. Competence or even brilliance in his chosen field—electronics, machining, nucleonics, internal combustion engines, or whatever—will not be enough. Some written expression of this knowledge, such as reports, letters, and so forth, will probably be required as part of his job. Therefore the clear and logical thinking that characterizes the good technician, engineer, or scientist should extend to what he writes; otherwise his influence and reputation will be severely limited.

In recent years, written technical instructions have been needed not only for the armed services but for manufacturers supplying their needs under federal specifications. In the mounting technical activities of today's booming atomic and space ages, the need for unmistakably clear instructions is much greater than ever.

But we must always remember that instructions themselves develop from the fully verified experiments and investigations of which individual reports are the first expression. So we have the two great divisions of technical writing: instructions and reports. (See Appendix, p. 161, for discussion).

The unalerted student who does not like to write may be tempted to regard a required writing course as just another passing discomfort of no permanent importance. Such a view would be a serious mistake, permitting faulty writing later that would be an obstacle to his success. If he writes well, however, he can acquire a kind of everyday halo; that is, he can be looked up to, admired, and depended on by management to express key situations and test results in terms of the written word. Sooner or later, the good writer achieves notice. He stands out, gets ahead in terms of status, salary, and opportunity. The author recently heard an engineer say of a colleague: "I don't think he's a better engineer than the rest of us, but he can write and so he's always asked to write up the department's monthly report. This one skill makes him useful to management."

The favorite method of teaching this new and highly diversified subject, technical writing, seems to be to paint a panorama of it. In most of the books on the subject, exercises are appended: "How many kinds of Type X writing are there?" "Summarize the job description of a technical writer." "List six different names used for technical reports." These exercises merely

1

reinforce one's memories of the text, which portrays the industrial writing field. There are books of this type that are worth their weight—but *the student's first requirement is to achieve skill in technical writing*. After that, it is indeed highly instructive to know how many different kinds of writing he may someday be doing or that someone is doing now.

Should a student, then, specialize at once? Should he achieve skill in a certain *kind* of technical writing, such as the report, or the "spec" (specification)? No, this is putting the cart before the horse. The type(s) of writing he is to do will evolve, broadly, from his chosen technology or science; narrowly, from the specific job he gets. It is easy to forget that truth. The author knew three writer-editors who, year after year, worked only on "TCTOs" (time compliance technical orders). Their extremely precise use of English was their essential tool, training for which would have been their best possible preparation. Yet how could they have known, as students, that they would be doing exactly this? Another employee has written nothing but specs for many years. Yet what guarantee is there that those who take expensive courses in nothing but "spec writing" will have such a career?

The student, here and now, cannot know which forms of writing he will be doing. To train for just one kind would be forbiddingly risky, since in his life he will probably hold more than one job and thus write more than one form of communication. His chosen vocation may even change over the years, perhaps radically. The wisest thing the student can do is to achieve a *general* competence, but *in* the technical writing field.

This field, however, covers everything from one-page interoffice memos to book-length manuals. Two essential writing skills, however, are involved throughout all this: organizing one's thoughts, then writing them into strong sentences. These skills *can* be applied in producing any type of technical writing—memo, report, manual, proposal—and *must* be applied if the work is to excel.

Is this not, then, just another English course ("no matter what they call it") copiously garnished, no doubt, with scientific terms? Not really.

Conventional freshman English is not for us. That type of course was fashioned long before technical writing became a roving specialty within various professions. Therefore the conventional freshman English course was never intended to train technical writers and cannot be blamed if it doesn't. Despite the general dislike of it by number-minded rather than word-minded persons, however, it has contributed invaluably by sharpening man's greatest social tool, language, and by representing, through literature, the world of art and culture. Conventional freshman English is priceless in its place, but it cannot be adapted to train technical writers and editors. For this it is helpful, but not helpful enough. There are two reasons.

1. Conventional freshman English courses offer training mostly in types of writing uncharacteristic of technical writing. They emphasize *narration* (e.g., first-person accounts of novel experiences and short stories) and *argumentation* (e.g., an assignment to prove or disprove that it is better to go to a university than to a small college; and debate-type questions.) Technical writing, on the contrary, emphasizes *exposition* (explanation is the "soul" of technical writing) and *description* (notice how many detailed situations, how much apparatus and instrumentation are described in technical writing). We are saying that the emphasis in the freshman rhetoric and English course is unsuitable for technical writing.

2. Conventional freshman English courses falsify the industrial writing situation; the student is characteristically presented with perhaps a dozen or more themes to write about, thus giving him deliberate latitude so as to enlist his interest. As little strain is put on him as possible so that his temperament and limited experience will be inspired by one of the themes.

But how different is writing in industry! Imagine a supervisor saying "Write a paper of about 5000 words on any one of the following themes."

Methods for Operating Visual Observation Equipment in Manned Orbital Laboratories

Comparative Times Required for Printing from Typewriter Copy and from Typeset Copy

A New Mathematical Code for Determining Gaseous Fusion Rates

Isotopic Power Systems for Space Missions

Guidance for Drafting Flip Charts

Determining Costs of Preparing Synthetic Salicylic Acid Compounds

It is necessary for the technical writer to work strictly to assignment. The difference between the two writing situations, once clearly seen, is startling. In the conventional freshman English course, whatever strikes the student's creative fancy is a major factor in the assignment. In industry, on the contrary, the writer does not as a rule create anything except an expression into English of the stubborn facts that he finds. He may be asked to make conclusions or recommendations at the end of his factual reporting—but he may also be asked not to, this being left to the judgment of management. In any event technical writing assignments are exacting and confining in the extreme. The writer is asked to ascertain the truth about a very definite situation. His "theme" is assigned by management, and his writing is controlled not by his own temperament, but by objective fact. He is limited by the situation, not freed by it to express his imagination. *How* he writes is important only to the extent that he expresses the facts clearly and concisely.

If he is asked to write a manual rather than a report, he again is strictly limited by the facts, which he must simply set down clearly and in proper sequence. This is true for nearly all other types of technical writing. The one thing the writer does not do is to "let himself go." A rich temperament gets in the way.

The preceding strictures are unfair to conventional freshman English courses only if we forget that we are, of necessity, talking in generalities.

We repeat that the essential skills needed for technical writing are the ability to organize one's thoughts or ideas and the ability to express them in strong sentences. Fortunately technical writing does not use a distinctive *kind* of sentence; nothing new need be learned about grammar, syntax, or rhetoric.

Here is a promise. Whoever works through this book conscientiously will come out on the other side of it forever different from the way he went in. If he dreaded the written word, he no longer will; and he will find a newly forged writing proficiency that will gladden and astonish him.

I

How to Organize Thoughts

Technical writing, like any deliberate activity, requires a plan. This means an outline. Would you believe this function can be mastered by using just four rules? Furthermore, if you outline with cards, it's easy to apply these rules—even pleasant! Comparing the topic outline with the sentence outline may change an amateur planner of writing into a professional one. You can bring your outlines to maximum efficiency by using the checklist on pp. 41–44. A power tool for writing is always at hand—the outline.

PLANNING, THE FIRST STEP TOWARD COMPETENCE

Generally speaking, the most difficult task for a technical writer is to plan for writing; that is, to organize his thoughts, ranking and ordering them according to importance. To make such work easier is the first skill to be acquired.

Planning is best done by outlining, which requires work because it is sheer logical thinking. (Conventional freshman English courses teach outlining, but not intensively enough.) If the duty is shirked, however, haphazard writing results. A writer who works without an outline is like a traveler who buys a ticket without knowing where he is going or how. It is, admittedly, easier to write without planning at all, or very much. But the result is either an inferior product or more work in rewriting than would have been needed if outlining had been done.

When we write without adequate plan, we almost always omit something essential. We do the writing, then we begin to feel a lack. But a lack of what? There is nothing we can look at to help us think of what we

5

need because we have put nothing down. We had no deliberate plan, no outline.

Any plan is made to be acted upon. A *faulty plan*—which certainly includes not writing one at all—will doubtlessly be expressed in faulty action and a *faulty product*. If you described how a Martian friend should dress himself in Earth fashion and left out belt or suspenders, would you be suprised if his trousers did not stay up?

A woman who gives a cooking recipe to a friend is morally obligated to include every ingredient in order to avoid some minor calamity.

Outlines are nothing but plans for writing (or for reading, if they are tables of contents), and incomplete plans lead unavoidably to unfinished products. What was not planned for in the beginning will not suddenly show up at the end. If the writer failed to think hard enough about his subject, two faults result.

First, a key topic may be ignored. Suppose, as a lighting specialist, you were asked to recommend either incandescent or fluorescent lighting for a new building. If you made your study without considering the probable cost of upkeep for either system, this would be an example of failing to think hard enough about the subject.

Second, a faulty plan often causes imbalance. Since the writer literally cannot see what he is doing very well, he may neglect to balance his outline by thinking parallel thoughts. If he wrote about the cost of upkeep for one lighting system but not for the other, this would be an example of imbalance.

The writer can often avoid imbalance simply by remembering to give equal treatment to equally important topics—something he can not very well know unless he has already coordinated them in an outline.

Rule 1. *Include Every Topic Required by the Subject*

Let us consider five examples of this first fault, namely, insufficient topics in the outline. Since technical writing can apply to anything technical, we are not surprised to note that these five examples refer to five different sciences or techniques. This indeed is a main point of this book: Skill in technical writing is to be sought because it can be applied to all of the exact and inexact sciences—and the arts as well!

Facility in outlining does not, of course, mean that the writer can somehow fake knowledge of his subject. Skill in manipulating ideas and expressing himself scarcely makes up for not knowing what to manipulate and express. This need for a store of facts to work with makes it important for us to choose five not too technically demanding examples that will make sense to the reader.

Machine shop theory and practice. Suppose a book on this subject starts with "I. Rules and Scales," under which appears "A. Definitions." This makes sense. Heading II is called "Precision and Semiprecision Tools" and begins similarly with "A. Definitions;" but Heading III is "The Lathe" and is not followed by "A. Definitions," nor is the next heading, "Milling Machines." More attention to balance and parallelism would have avoided this discrepancy.

We do not need to understand machining to see that this textbook has been written from an incomplete outline. The lack of definitions for basic instruments may impede a conscientious student who wants to start his study of machining by grasping exactly what each of the operations is and does. When he cannot find out, he begins to lose faith in what the book will give him.

Industrial engineering. You are asked to make a savings analysis of bulk freon purchase. Should your company continue its present method of buying this industrial gas in 55-gallon drums or should it install a single below-ground storage tank of 3,500-gallon capacity, the freon to be delivered in bulk by tank truck? You carefully contrast the present method with the proposed method, in regard to freon cost by drum and in bulk, handling cost, capital investment required for any change, and so on. But you fail to think of the material loss over a period of time, in each of the two methods.

Your supervisor points out, after you have turned in this report that required a month's hard work, that freon leaves a "heel" in each drum averaging 1.5 gallons per drum. This equals a freon loss of 12,355 pounds per year, whereas the tank storage method leaves no heels. By failing to think of this topic for investigation, you have ignored a loss, under the old method, of $6,795.25 per year. Whether or not this changes your recommendation, you have made an incomplete study for the company because of a failure to consider the assigned subject very carefully.

The gasoline engine. This time, let us begin by showing the outline fragment itself. Suppose a class had been studying the ordinary automotive gasoline engine and was asked to outline its knowledge of the topic, represented by I, as follows:

    I. Strokes in Operation of a Gasoline Engine
        A. Intake
            1. Volume Increase
            2. Pressure Decrease
            3. Entrance Into Cylinder of Gas Vapor and Air
            4. Closing of Intake Valve

     B. Compression
       1. Volume Decrease
       2. Temperature Increase
     C. Power
       1. Volume Increase
       2. Pressure Decrease
       3. Opening of Exhaust Valve
       4. Exit From Cylinder of Part of Gas Vapor and Air

Let us say that you have been asked to criticize this outline fragment. What do you notice?

First, the fourth stroke, "D. Exhaust," has been left out; this alone makes the outline inadequate because it is incomplete.

Second, the explosion leading to the power stroke has not been mentioned. We should have something like "B.3. Ignition of mixture by spark."

Third, only B.2 mentions temperature. Should a temperature increase or decrease be mentioned in each of the four steps? Maybe not. This will depend on our purpose and perhaps on how detailed we want the outline to be.

Fourth, and similarly, should the increase or decrease of volume and pressure be mentioned for each of the four strokes? This, too, must be decided on in light of one's purpose.

Notice how much easier it is to work with your ideas in outline. You can see what you have. You can more easily spot the empty or inconsistent places. Otherwise, after writing up the report, you are able only to *feel* that you lack something, without knowing what. To discover just what this is requires that you start all over again, this time by outlining. Why carry the topics in your head as you write? That is doing it the hard way.

The time taken to prepare an adequate outline should never be deplored, for outlining is a perfect example of the adage "The longest way 'round is the shortest way home."

Printing. Suppose you were planning to compare the two main printing methods of letterpress and photo-offset. The first rather skimpy outline might be something like this:

     I. Description of Process
       A. Letterpress
       B. Photo-offset
    II. Relative Merits of Each Method
       A. Quality of Product
       B. Cost
   III. Conclusions

Using either your knowledge of this subject or the vigilant common sense needed to write anything, what topic would you say is missing and most needed? Think about it before reading further. As you have noticed, we follow the questions in this book with their answers so that the student can check his own.

The topic missing in the above outline fragment is "Time Required to Print," certainly a leading consideration.

Magnetism. Let us give in a slightly different way the last example of the first great fault in outlining, namely, not including enough topics. The following are the main topics (first-order headings) from two different chapter outlines, A and B.

A
1. Natural and Artificial Magnets
2. About the Lodestone
3. The Earth as a Magnet
4. Artificial Magnets
5. What Magnetism Is
6. What the Magnetic Field Is
7. Magnetic Flux and Magnetic Density
8. The Oscillation of a Compass Needle
9. How Magnetism Acts Through Various Substances
10. Magnetic Attraction and Repulsion

B
1. Natural and Permanent Magnets
2. Coulomb's Law of Force
3. Magnetizing by Induction
4. Intensity of Magnetic Field
5. Mapping the Magnetic Field
6. Field Intensity Around a Magnetic Pole
7. A Theory of Magnetism

By comparison one notices that B does not feature the earth as a magnet (A.3), nor does it tell how magnetism acts through various substances (A.9). Both are interesting topics.

Both outlines fail to include an explanation, however simple, of how magnetism is measured, which would involve the subject of magnetic moment. Only the student who has studied magnetism would be likely to notice this one, and he could tell us, further, whether B.2 and B.3 are important enough to have been included in A. (We would say they are.)

Looking at outlines like this makes us realize that the technical writer's responsibility includes envisaging *all* of his given subject. This sounds easy, but it isn't.

The following is the first self-administered test. Please answer (in ink, on scratch paper), the 10 questions, which are based on the text so far; this includes "The Special Needs of Technical Writers," which starts on p. 1. To verify your correct recalls or to find answers, refer to the text pages mentioned at the end of the questions. This process will tend to familiarize you with the text and provide valuable drill in recognizing topics. The learning process, which is surprisingly complicated and still mysterious, will be aided thereby. As a final check, turn to Exercise 1 as answered in the Appendix, p. 163.

Exercise 1. *Include Every Topic Required by the Subject*

1. Why should a person who plans a technical career be able to write well? (p. 1).[1]

2. What reasons were given for technical writing being so different from the conventional freshman English? (p. 2).

3. Why does the best preparation for technical writing foster a general competence in that field? (p. 2).

4. Should not skill in technical writing help to explain *any* scientific procedure? Defend your answer (p. 4).

5. Is it true that the most difficult task for a technical writer is to plan for writing? Explain your answer (p. 5).

6. All faulty plans, applied, result in faulty performances. Some outlines are faulty plans, applied. Therefore. . . .
Please finish the syllogism. It is a valid argument; if its first two statements (premises) are true, so is the conclusion. This latter may not be very surprising, but it is a reminder that the faulty planner has only himself to blame if his writing is confusing and a failure (p. 6).

7. What are the two main reasons for failing to include enough in a writing plan? (p. 6).

8. Refer to text, then state at least one needed topic that was left out of each of the five technical idea-groupings or outlines given (pp. 7–9).

9. Compose a short outline of your own from a technical subject of your choice. Based on your own knowledge, try to add a topic to this outline. Can you see how you might add further topics to it, through study? If so, what precisely would you study?

[1] A text reference may extend beyond the page mentioned.

10. State how much outlining you have done so far in planning your own writing, whether technical or not.

Rule 2. *Exclude Every Topic Not Required by the Subject*

This is not merely Rule 1 differently expressed; it is a separate rule. This is proved by reflecting that each rule could be satisfied without satisfying the other. One could include every topic required by a subject, yet fail to exclude every topic *not* required. Similarly, one could exclude every topic *not* required by a subject yet fail to include every topic required.

Let us consider this second type of error that is made when we write without planning. We include too much, namely, topics that do not belong. We are not referring to needless words in sentences—this will be discussed later—but to needless ideas in written compositions, whether these be reports, manuals, proposals, or anything else. These unneeded topics often creep into a writer's work even when he is deliberately and carefully outlining. Fortunately they can be discovered by careful inspection.

It will help to illustrate the second type of fault if we reuse our examples. So now we are listing, for our Martian friend, items needed for him to be dressed in Earth fashion. If we tell him to wear a cap with a feather in it, this will be harmless and piquant, but it won't be characteristic.

Or you are asked, as a lighting specialist, to recommend either incandescent or fluorescent lighting for a new building. If your oral or written report includes a detailed recommendation for soundproofing, your report as well as you will be viewed with doubt.

Or, in the textbook on neurology, say you are writing about the sense of sight. While researching you become absorbed in the psychology of color and you take copious notes on it—only to realize, when writing, that this absorbing topic just doesn't belong in the book. Psychology is not neurology.

Or, in doing research on the sense of hearing, you might be surprised to learn that what a civilized person regards as the "natural" musical scale is not a fact of nature but of culture. The ancient Greeks had a different scale; they made a succession of tones different from ours when they sang or played. The modern Chinese do too; they have their own scale, which seems as natural to them as ours does to us. You would not want to waste this research when you came to write about hearing; but you would have to, since the subject doesn't belong to neurology.

To keep irrelevant topics from creeping into writing plans, it helps to remember how this happens. There are several reasons: (1) We do much work on something that interests us, only to realize later that it doesn't belong; our enthusiasm has led us astray. (2) A subject is *so close* to our topic that it seems, at first, to belong to it. (3) After researching, we decide that our original subject is too broad; narrowing it down, we find some of the work we have done is useless.

Let us indicate how unneeded notions could intrude into each of our five chosen instances.

For "Machine Shop Theory And Practice," under "Rules And Scales," a person might find he had unwittingly collected interesting material on why the metric system, instead of the "pound-inch-quart" system, should be universally adopted. Such material probably does not belong in the book unless it immediately concerns machinists. One's best judgment is often required to exclude topics. Ruthlessness becomes a virtue.

In regard to the "Industrial Engineering" example of whether to purchase freon in bulk, suppose you had found an authoritative article on how to construct underground storage tanks. Your satisfaction would be understandable because such knowledge could increase your value to management should it decide to build the tank. But your assignment did not include this. Besides, management probably knows that the storage tank can be built; otherwise your analysis would not have been asked for.

In the "Strokes in Operation of a Gasoline Engine" sample outline, one would not expect to see topics dealing with the carburetor. Although vital to engine operation, the carburetor's work does not consist of the piston strokes.

In regard to comparing letterpress and photo-offset methods under the printing example, you might be tempted to include various reflections on editing. Or you might want to explain how the linotype machine, which had been necessary in the use of the letterpress method, can also be used in photo-offset work. But chances are that these additions to an already highly technical and complicated subject would only confuse the reader— and it is for him that all our writing is done.

In "Magnetism," under the subject of artificial magnets, there would properly be mention or an explanation of the ship's compass. But colorful anecdotes on how mariners have depended on the compass through the centuries and how much civilization owes to it would be out of place in a physics textbook. In considering the nature of magnetism and the magnetic field, detailed biographical accounts, however interesting, of the contributions of Wilhelm Weber and Sir Oliver Lodge would be irrelevant.

The publications chief will have his own superiors to report to. He will find them—and thus himself—very sensitive to company policies. He probably will have to conform to most of these and must learn when to pass his problems on to superiors, and when not to.

Does the paragraph you just read irritate you? Good: it does not belong on the page. It was deliberately inserted out of context to show what happens when you write up some topic that should never have been in your outline. Unless you see clearly that the topic belongs, it should be left out—even if it *almost* belongs. But how can you decide this?

There is often a simple, effective way to decide whether a particular idea or topic belongs to the subject about which you are planning to write: Before starting to research or outline, form if possible a *limiting sentence* (also termed "topic sentence") from your subject or thesis. Write it down on a piece of paper so that you can refer to it. It should, ideally, be a "declarative" sentence, not a question or instruction. Then discard each idea or topic that is not contained by your limiting sentence. Let us see how this works.

Here are some limiting sentences (or "themes").

1. Rearranging the shop layout according to Plan C will cost $50,000 but will pay for itself in two years.
2. My investigation shows that in Department A the inferior product Z is due to adding material X to material Y and that, if X is left out, better quality will result and costs will be lowered.
3. Setting up complete facilities for in-plant printing will improve the quality of our published products and will greatly decrease the time needed to print.
4. Sodium is preferable to organic fluid as a coolant for nuclear fission.
5. Routings taken from expense reports and sales records for the last two years show that Sales Territory 12 can be abandoned beyond the western boundary of Illinois, resulting in a saving in expenses and a concentrated working of the larger accounts.
6. Tests indicate that all statistical information now required from the five main departments of the Robertson plant can be gathered by the installation and use of one unit of the ABC tabulating computer.

The best way to exclude needless topics, then, is to start organizing material by carefully formulating and stating your limiting sentence. In the light of such a key sentence, you examine each topic and reject it if it does not belong; for example, in writing about the first of these limiting sentences, "Rearranging the shop layout . . . ," could you use your knowledge of other plans, say A and B? No; the topic is Plan C, not Plans A and B. *Your rereading of the limiting sentence reminds you of this.* Or suppose, in connection with limiting sentence 2, "My investigation

shows . . . ," you knew all about the inferior product M. Would this belong in your report? Hardly; another inferior product, Z, is being analyzed.
Reading about M would only confuse the issue.

Once in a while a single topic affects the limiting sentence itself, instead
of the reverse. That is, you may find a topic that definitely belongs to
what you want to write about but is not included in the limiting sentence.
In this case, the limiting sentence is at fault and must be amended. For
example, under 3, suppose you had reason to believe you could indeed
achieve better quality through in-plant printing but that, for various reasons,
such printing would take just as long as "vendor" (outside firm) printing.
If this were true, you could still keep most of the limiting sentence, but
would have to delete "and will greatly decrease the time needed to print."

Having a good limiting sentence is an ideal and, when unattainable, one
must construct an outline without it; for this desired sentence may express
what was not known at first but was learned during or after research.
Thus it would be a conclusion, arrived at after all work had been done
except planning and writing the report.

But what will help in the beginning if one does not have a limiting
sentence? The answer is, something as close to it as one can devise. It
might be a question such as "should the shop layout be arranged according
to Plan C?" Or it might be an instruction, "Compare sodium with organic
fluid as a coolant for nuclear fission."

A sentence says something, makes a statement about, asserts a connection
with, a subject. (The limiting sentences of 1, 2, 3, 4, 5, and 6 are examples.)
Many times, however, especially in "R and D" (research and development)
work, one does not know what statement he will be able to make *about*
a subject and must therefore be content with *only that* subject; for example,
"Plans to Harness River X," "Advanced Use of Plastics in Heart Surgery,"
or "Alleged Waste in Department K."

There are also times when only a good pro-and-con discussion of a subject
is desired. This is the "balanced report," which describes two contrasting
situations without making recommendations.

In all these cases, the limiting sentence will be very hard, if not impossible,
to come by and can therefore be used only when technical writing situations
permit.

Form a limiting sentence if you can, but do not feel lost if you lack
one. A book, for example, has a table of contents, and this is an outline.
But rarely can such an outline be adequately expressed in a single sentence.
A textbook on physics cannot be expressed in a sentence because the purpose
is to discuss the entire subject, which is composed of many parts. If at
work you were asked to write a manual, this means you would be writing

a book of directions; its table of contents would hardly be expressible in a single sentence.

Of course a subject guides the writer much better if it has a predicate. Consider "1" again. If one is deciding how much of the data collected belongs to a report and how much does not, would it not be more helpful to be able to refer to the sentence, "Rearranging the shop layout according to Plan C will cost $50,000 but will pay for itself in two years," rather than to the topic, "Rearranging the Shop Layout"? Or, (4) to test one's collected data against the statement, "Sodium is preferable to organic fluid as a coolant for nuclear fission," instead of against the phrase, "Sodium as a coolant for nuclear fission"?

Here is a practice that greatly helps one engineer in writing his reports. He first writes an abstract (discussed in Chapter IV, "Turning Out a Written Report") ; he then determines whether each topic he is thinking of dealing with has a place in the abstract. Since abstracts are usually written last, this practice would not normally be possible. But if you do know enough about your subject and what you are going to write to be able to write at least a preliminary abstract, it should be an even better guide than a limiting sentence.

Exercise 2. *Exclude Every Topic Not Required by the Subject*

Please answer, as in the first exercise, the following 10 questions. Directions are on p. 10. Suggested answers are in the Appendix, p. 164.

1. How many unjustified reasons were mentioned for allowing irrelevant topics to creep into writing plans? What are these reasons? (pp. 11, 13).

2. Do irrelevant topics get into your own writing? If so, is there any one explanation or are there several? In future work try to note your tendencies in order to counteract them.

3. What happens to one's writing product when he has allowed an irrelevant topic, or several, to remain in his outline? Describe the effect of such unprofessional planning on the reader (p. 13).

4. What is the best way to exclude needless topics? (p. 13).

5. How should one use the limiting sentence? (p. 13).

6. Has one ever the occasion, as he outlines, to change the limiting sentence itself? If so, describe the situation (p. 14).

7. Explain what is meant by saying that the limiting sentence is an ideal in technical writing (p. 14).

8. Do you think the limiting sentence itself can be part of the final writing product or does it merely help with the writing?

9. Why would a text book not likely be written around a limiting sentence? (p. 14).

10. "Of course a subject guides much better if it has a predicate." What does this sentence mean? (p. 15).

Rule 3. *Working from the Top Down, Divide Each Topic into All Its Subordinates*

The nature of the outline, which is a system of specially related parts of decreasing importance, makes this a sure-fire rule. The "top" is of course the subject or chief topic itself. If this did not "divide," that is, have parts, there could be no outline. Such parts of a topic, its (immediate) subordinates, must be recognized by analyzing it. There are always at least two of these parts on each level of outlining, simply because when something is divided at least two parts result. There may be *more* than two parts, however, which gives the "all" in Rule 3 its force.

All the parts that immediately result from dividing the same topic are "coordinates," or equals. At the various levels during the work of outlining, attention often shifts from subordinating to coordinating. When one finds a subordinate, it may be easier to start looking for *its* coordinates than to search for other *subordinates* of the parent topic. In any event *each* topic in an outline is *both* a subordinate of its parent topic, or "superordinate," and a coordinate of its fellow coordinate(s).

When all coordinates of any one group at any one level are discovered, continuance or extension of the outline requires that at least one of these coordinates, in turn, be divided into *its* subordinates. For the sake of thoroughness, every member of each group of coordinates is tested to see if it will father any subordinates. Whenever it does we have a lower order of topics to analyze, in the same way as before.

So, vertical and horizontal movements of thought can tend to alternate. There is no need for this psychological fact to be confusing. Common sense can still order the process at every step. One begins by dividing (finding the subordinates of a topic), finishing each such act before starting the next one. The resultant subordinates are related to one another, on any one level, as coordinates. Each coordinate is then to be tested to see if it has next-lower subordinates of its own.

Let us rethink the situation because a grasp of this most difficult part of outlining will make the organizing of our thinking a pleasure. To subordinate and coordinate each topic correctly, we must find, except for the subject itself, (1) the one and only topic to which it must be *subordinated*—its superordinate—and (2) at least one other topic with which it must be *co*ordinated. These are the only two conditions that must be met by every topic except the subject. (The main subject would not have a *co*ordinate.)

To do this work for every topic in an outline is not as difficult as it sounds. The relationships are already there in abundance, ready to be recognized. Furthermore, the topics are "correlatives," that is, each is related to at least two other topics. Whenever any of these relationships is discovered, the associated relationships are thereby discovered or are more likely to be. For example, when the outliner spots a subordinate of a topic, he has spotted that subordinate's only superordinate. When he finds one coordinate of a topic, he may then more easily find another if it exists. Every outline discovery helps to make others.

It is as though a group of people were thrown together in an emergency (let us say they all have to make outlines if they are to communicate!) and had to organize themselves at once so they could act efficiently and without misunderstandings. This would require that each person (topic) have an immediate superior (superordinate), except the head of the group (the subject itself). Those in the same group having the same superior would be equals (coordinates).

In an outline, two or more coordinate topics always have the same superordinate. In fact, they are coordinate or equal in our eyes only because their immediate superior is the same. They are equal (to each other) because both are subordinate to the same third topic. Their being alike in that precise respect is what makes them members of the same class. "Monday" and "Tuesday" are coordinate as class members of their superordinate, "days of the week." We would not be apt to wonder if the binomial theorem is equal to the Supreme Court or if baseball is as important as Pittsburgh; these pairs of topics have no reasonable superordinates and are not, therefore, outline material.

Subordination discoveries are necessarily made first since outlining depends upon this dividing. They are also easier to make; to subordinate one topic to another requires perception of an *immediate* relation, whereas to coordinate two or more topics requires perception of a *mediate* relation, any two coordinates being related mediately only because each is related immediately to its superordinate.

Before proceeding, let us clear away some underbrush. Does not the writer who has finished his research on a subject already have enough of an outline in his mind—an adequate "informal" outline? Let us discuss the reasons for him to (mistakenly) believe that he does.

Admittedly we do not first gather topics and then classify them all into a finished outline. We do some of the outlining as we collect data, adding or rejecting topics, thinking out dimly or clearly certain connections or lacks of connection among the ideas we encounter. As we read, those parts of our subject indicated in the text itself as subordinate or coordinate may not even be suitable as topics in our own plan. But much acceptable material will probably be gleaned from research.

The topics found, however, including one's own first hasty subordinating and coordinating decisions, are neither exact enough nor numerous enough. The informal, unexpressed, tentative, born-of-research outlining must be followed up by the formal, definitive, carefully stated-in-writing, precomposition outlining. That some early, fragmentary, semiconscious outlining has occurred must never beguile the writer into thinking he does not need the later explicit type.

Such early semi-outlining is, however, aided by another factor: the technical writer, unlike the college theme writer, usually has part of an outline imposed upon him from the very first. He knows he is to write, for example, a company report of a certain type, which will have certain elements that must appear, as for example an abstract, an introduction, a summary, or a recommendation. These, being parts, must be parts of the outline. But such common elements, since they are the same for all reports of that type, can never differentiate each report. The work of adding all the individualizing topics to the outline, so as to characterize just *that* report, must still be done.

What results from a failure to observe the third rule for outlining ("working from the top down, divide each topic into all its subordinates") is more upsetting to readers than a failure to observe the first or second rule. The writer can neglect to include something important or can add something unimportant without alienating his readers. But if he fails to organize correctly what he does present, readers may be disaffected at once.

Unfortunately, a writer may dislike being *deliberate* about all the subordinating and coordinating decisions necessary. It is easier, he tells himself, to make them informally "in his head"—which usually means not outlining correctly if at all. Writers who will not do this work deliberately when it should be done make it easier for themselves initially, but harder for themselves and their readers later on.

This failure to outline adequately is due mostly to a lack of practicing the skill. What an outline is may have been explained in an English course; but the vital importance of the activity for technical writing has, understandably, been slighted. This underexposure must be remedied. Those who begin outlining are often surprised to find that this activity appeals powerfully to their *workmanlike* tendencies.

Let us do some outlining. The main subject divides into *first-order headings,* whereas *any* topic that is written about, or that appears anywhere in the outline, is a *heading*. It is usually preceded by a symbol that indicates its importance relative to other headings. Thus "I" means that the topic that follows has been classified as a first-order heading. These procedures are explained in "Classifying Devices of the Outline," p. 33.

To repeat, any divisions of a subject are all *sub*ordinate to it, but they all *co*ordinate with each other.

The first-order headings of many outlines divide into second-order headings, some of these into third-order headings, and so forth. Let us work with a few examples.

Suppose we have finished our research with a great many notes that include these key topics: How Heat is Produced, Electric Currents, How Heat is Measured, Chemical Action, Heat, How Heat is Transmitted, and Mechanical Work. Seven topics? Yes, but we are investigating heat, which is therefore our subject, and the other six topics bear *some* subordinate relation to it. We say "some" because we must not assume that all six are first-order headings. We are not yet sure of that. Some (or maybe other topics we have not yet thought of) *must* be, however, because an outline cannot jump from its subject to its *second*-order headings, without having first-order headings. Let us assume, reasonably enough, that we have the first-order headings in our collection. How do we single them out?

We look for those that seem directly or *immediately* subordinate to the subject. These are the three "How . . ." topics. If we stop outlining at this point, assuming that the other three topics are also immediately related to the subject (which we said was a risky assumption, because untested), our outline will look like this.

HEAT

    I. How Heat is Produced
   II. Electric Currents
  III. How Heat is Measured
  IV. Chemical Action
   V. How Heat is Transmitted
  VI. Mechanical Work

We have researched a long time and we want to start writing, but our topics are still a hodgepodge. If we wrote them up, we would be going from heat to electricity, back to heat, then to chemistry, to heat again, and finally to mechanics. This is not orderly, so we must conclude that more outlining is to be done.

Since this is a typical example of being "stuck," it is the place to ask how we become unstuck. First we review what we have. Certainly "heat" is the subject, and its division into the I, III, and V headings seems to make sense. If the outline is to continue, we must find a place to do another act of dividing. We have three possible places—I, III, and V. Since II, IV, and VI do not seem to belong directly under Heat, maybe one or more of them belong under I, III, or V.

Let us try I. To ask if I divides partly into II is to ask if II is subordinate to I—that is, to ask if heat can be produced by electric currents. We remember at once that sending electricity through carefully planned wiring resistance makes possible electric heating. Using electricity is one of the methods for producing heat. So "Electric Currents" seems to be subordinate to "How Heat is Produced." This would make "Electric Currents" a second-order heading, preceded by a capital letter such as "A."

When in outlining we find a solution to a problem, we try to apply that solution to other problems—of which we now have two left. If II is the likely subordinate of I, maybe IV and VI are also subordinates of I.

As we did with I, we translate into what we are really asking. In the case of IV, it is whether heat is produced by chemical action. We need remember only that oxidation is a chemical action and that the burning (a type of oxidation) of wood, coal, and so forth, produces heat. Thus IV also belongs under I. Does VI too? Is heat produced by mechanical work? Yes; we remember the heat of friction.

We have been lucky to discover three subordinate relationships without much effort, using only one of the three superordinates we had to consider (I, instead of III and V too). But it is a good idea to confirm such discoveries by checking around a bit. How? By assuming that II is *not* subordinate to I, so that we can see how feasible another superordinate for II would be.

Let us try III as that superordinate. Would it make sense to place "Electric Currents" under "How Heat Is Measured"? No; one characteristically measures heat by thermometers rather than by electric currents.

Now let us check further and try V as the superordinate of II. That is, do electric currents transmit heat? No; they can be used to cause it but not to transmit it.

So our outline becomes the following.

HEAT
I. How Heat is Produced
   A. Electric Currents
   B. Chemical Action
   C. Mechanical Work
II. How Heat is Measured
III. How Heat is Transmitted

Note that since the topics following A, B, and C are subordinated to a topic bearing a number, namely I, they carry letters. An outline should alternate the use of numbers and letters.

In this general way, then, we internally relate a set of topics that already satisfy our first two rules (1. Include Every Topic Required by the Subject, and 2. Exclude Every Topic Not Required by the Subject).

Most outlining tasks are more difficult than this one, but the stages are similar. The order of discovery will vary somewhat with different persons and from one outline to the next with the same person. But everyone uses subordinating and coordinating topics because an outline comes into being only by the division of topics.

Let us outline a little more. Would you say these topics belong together: (a) Automobiles, (b) Modern Physics in Everyday Life, (c) Medical Instruments? The second *sounds* like a subject, does it not? But are a and c subordinates of it? To answer, we try to see how each can be related to b. Yes, the automobile is an example of physics in everyday life (expansion of gases according to Charles' law, and so forth), as are many medical instruments (selective penetration of living substance by electrons of X-ray machines, and so on).

Then a and b are coordinates? Yes. Although one does not usually group automobiles and medical instruments together, both are examples of applied physics.

To use another analogy, topics positioned in outlines are like individuals who suddenly discover at least two close relatives and maybe more. Why at least two? Because when we see how to subordinate Topic X we see what its superordinate is (its "father" in meaning). That makes one new relative. But we also see at least one of its coordinates (a "sibling," i.e., a brother or sister). Remember that if we divide anything, we get at least two parts. The superordinate of Topic X has been divided already in order to yield X. But there would be at least one other part of the superordinate left, say Topic Y. This would be the sibling of Topic X, which thus achieved at least two new "relatives" from the relating act, namely, a parent and a sibling.

Not every topic has to be divided into two or more parts. If this were the case, an outlining task would never end. Some topics do not fall naturally into parts and can be adequately treated where they occur. On the other hand, parts themselves can have parts. The shape and articulation of each outline depends on its subject, on how its parts divide up for treatment, and on the writer's purpose.

Let us say we have just figured out the costs of irrigating Old Macdonald's farm; it is doubtful, however, that the bank will lend the necessary funds. The report from which this will be decided will have to be crystal clear. Does a description of the farm have equal status with the costs of irrigation? Probably, for these are based on certain farm characteristics that should be mentioned as justifications of the expenses. So it looks as if we will have two first-order headings.

Our research, including several talks with salesmen, has left us with many figures and these jumbled topics: Connections, Materials, Tertiary Pumps, Installation, Main Pump, Operation, Costs of Irrigation, Secondary Pumps, Description of Farm, Sprinklers, Piping, Maintenance, Pumps, Plumbing.

Two of our first-order headings are there: "Costs of Irrigation," and "Description of Farm." Are we sure there are no more? We apply our method: are any of the other topics directly subordinate to the main subject, "Cost of Irrigating Old Macdonald's Farm"? How about the first topic in the list, "Connections"? No. That is a type of cost, but it seems subordinate in some way to "Costs of Irrigation," which is *already* a first-order heading under its superordinate, the main subject. Is "Maintenance" a first-order heading? No. That is a cost too and, like "Connections," is closer to "Costs of Irrigation" than to the subject itself. All the other topics, in fact, seem to be costs of some sort—which means that the other first-order heading, "Description of Farm," *will not* have subheadings. So our first subordination operation indicates that we will have only two first-order headings.

But is it just a matter, perhaps, of classifying "Pumps"? No. We are dismayed to notice "Sprinklers" in the list. These certainly are not pumps. And what is "Materials"?

Frankly, we felt good about outlining until now. What should we do when confused? Well, only this has happened: we realize that the outline will be more complicated than we had hoped. It is very easy to rescue ourselves, however, by applying an invaluable rule of thumb. We drop everything, go back and do what we should have kept doing—look for the subordinates of the topic that we are *most interested* in dividing. What topic is that? It has to be "Costs of Irrigation," since this clearly seems to comprise several kinds of costs and the other first-order heading that we are sure of, "Description of Farm," does not seem to have any parts among the topics. We have already observed that some topics, for example "Tertiary Pumps," are not *immediately* under "Costs of Irrigation." We want to discover the topics that *are*—they will be the second-order headings that we now need in order to continue outlining. We bank, as usual, on recognizing the immediacy of their relation to their subordinate, "Costs of Irrigation."

Outlining progresses by our testing one small new theory after the other. At this point, the theory is that " 'Costs of Irrigation' has topics (they would be second-order headings) subordinate to it that we can recognize." We run our eye over the list again. Yes, "Materials," "Installation," "Operation," and "Maintenance" are four likely costs of irrigation. All the different pumps would fall handily under the first of the presumed second-order headings, "Materials"; so would "Sprinklers." So far, the upcoming facts support our first theory.

Would remaining topics fall under the other three candidates for second-order headings—"Installation," "Operation," or "Maintenance"? No; then we remember that *a topic does not have to divide*. But if "Installation," "Operation," and "Maintenance" do not divide, in one way or another the remaining topics would all have to fall under "Materials." Do they? Let us see.

All the specific kinds of pumps would fall under "Pumps," (which, we said, falls under "Materials"). This leaves us with only four topics to place: "Connections," "Sprinklers," "Piping," and "Plumbing." Yes, these are all materials that would have to be bought for an irrigation system. Do they, then, satisfy our test, that we can see each one subordinated to "Materials"? "Sprinklers" and "Plumbing," yes. But we observe that the materials of plumbing *are* "Connections" and "Piping," which seem therefore more closely subordinate to plumbing than to anything else. So let us make "Plumbing" the superordinate.

We have found lodging for all our topics; let us set the outline down.

COST OF IRRIGATING OLD MACDONALD'S FARM

I. Description of Farm
II. Costs of Irrigation
   A. Materials
      1. Pumps
         a. Main Pump
         b. Secondary Pumps
         c. Tertiary Pumps
      2. Plumbing
         a. Piping
         b. Connections
      3. Sprinklers
   B. Installation
   C. Operation
   D. Maintenance

Considered descriptively, there are two first-order headings, four second-order headings under the second first-order heading, three third-order headings under the first second-order heading, three fourth-order headings under the first of the third-order headings, and two fourth-order headings under the second of the third-order headings.

Let us generalize a bit and review. An outline is a collection of topics, or sometimes statements, arranged to permit unified treatment of a subject. It is a carefully organized structure of leading, interrelated thoughts that have been collected through the researching and/or previous knowledge of the writer. An outline is, then, a kind of blueprint or X-rayed skeleton that shows the plan for a piece of writing. The topics used are also called headings because each may head up a paragraph or more that, when written, will flesh out the skeleton of the outline.

The advantage of knowing what we do when outlining is that we can then keep working correctly and without discouragement. At the beginning our topics are always quite jumbled, but we proceed confidently, one step at a time, and apply our rule over and over again—"Working from the Top Down, Divide Each Topic into All Its Subordinates."

Any topic X can be *sub*ordinated to only one (other) topic; that is, it can have but one superior or superordinate. Thus there is only one topic for "Plumbing" to be subordinate to, namely, "Materials." But any topic X can *have* more than one subordinate; thus "Plumbing" itself has two, namely, "Piping" and "Connections."

A topic may be *co*ordinated with several other topics; we have just seen "Materials" coordinated with the three other second-order headings, "Installation," "Operation," and "Maintenance." I and II are coordinates; so are 1, 2, and 3, a, b, and c under 1, and a and b under 2.

Every subordinate in an outline requires its superordinate to explain it, to make clear what it is doing there. Thus the superordinate "Plumbing" is needed to meaningfully relate its two subordinates "Piping" and "Connections."

How are outlining and classifying related? Outlining is a specialized kind of classifying, done either to help in writing the topics outlined or to comprehend these topics rapidly, as in a table of contents. Classifying in general is more dependent on preexistent facts than is outlining. One classifies what one finds; one outlines what meets one's purpose.

An imperfectly outlined piece of writing means that the writer neglected to make all the relating decisions that it was his duty to make. As a result the annoyed reader will wonder what these relations are and try to do for himself the writer's work; or he will just be confused. In either case he will be prevented from concentrating only on the intended meanings. A writer should correctly outline first so that he will (1) clearly see the relative importance of his topics and thus be able to give them their corresponding emphasis and rightful wordage in the finished product and (2) have before him a plan to treat all his topics in their proper places, thus giving his readers a sense of continuity and order.

By outlining we also catch many mistakes that we would otherwise make.

Exercise 3. *Working from the Top Down, Divide Each Topic into All Its Subordinates*

Please answer the following questions, as directed on p. 10. Then read Appendix, p. 165.

1. What is meant by saying that vertical and horizontal movements of thought tend to alternate in outlining? (p. 16).

2. What are correlatives? How do they make outlining easier? (p. 17).

3. What conditions must be fulfilled for topics to be coordinates? (p. 17).

4. Are the following two statements (not directly from the text) true? Defend your decision.

> If the explanation of two topics being related is found by thinking only of *them*, one is subordinate to the other. But if such explanation is found by thinking of a *third* topic, these two are themselves coordinates, both being subordinate to that third topic (p. 17).

5. In outlining, what two types of logical acts do we keep making? Which comes first? What is the difference between immediate and mediate relations? (p. 17).

6. What does an informal outline comprise? Why is it inadequate? (p. 17).

7. Do you agree that "failure of the writer to observe our third rule for outlining is more upsetting to the reader than failure to observe the first or second rule"? Can you state why this is so or why you think it is not? (p. 18).

8. How did we proceed when we "got stuck" in outlining "Heat"? (p. 19).

9. In outlining, what is meant by saying one tries to use a solution to solve more than one problem? What part of Rule 3 might such a practice fulfill? (p. 20).

10. Why does each topic, when correctly placed in an outline, always have at least two "close relatives"? (p. 16).

11. Why must there be at least two coordinates subordinated to a subject or topic when it is divided? Must every topic be divided into at least two subordinates? (p. 16).

12. Are you able to identify, characterize, and describe your own processes as you subordinate and coordinate topics? Whatever you can introspect on this subject, as you work, will make outlining easier.

13. Is every topic belonging to an outline the subject of both a subordinating and a coordinating decision? Explain your answer (p. 16).

14. What is meant by saying that "outlining progresses by testing one small theory after the other?" (p. 23).

15. Compare and contrast a "topic" with a "heading" (p. 18). How are "outlining" and "classifying" related? (p. 24).

Rule 4. *Order Each Group of Coordinates Properly*

The last of the four rules for outlining is the easiest to apply. When we were looking for second-order headings under "Costs of Irrigation," the "loose," unarranged topics in the MacDonald Farm list happened to be already in the correct chronological order; that is, we would have to obtain the materials before we could install them, and so forth. "Materials," "Installation," "Operation," and "Maintenance" were our A, B, C, and D.

But not all serial orderings are as simple. A previous outline (p. 20). resulted in the following:

HEAT
   I. How Heat is Produced
      A. Electric Currents
      B. Chemical Action
      C. Mechanical Work
  II. How Heat is Measured
 III. How Heat is Transmitted

The decisions to place A, B, and C under I did not give us their order, nor of course tell us how I, II, and III should be ordered. The ordering, that is, the proper arranging of a group of coordinates, is always a separate act of judgment.

Every group of coordinates should be free of "cross-division." "Cross-division" is the use of more than one criterion in choosing the subordinates of a heading; and it renders a subject unclear. To return to our irrigation example:

A. Materials
   1. Pumps
      a. Main Pumps
      b. Secondary Pumps
      c. Tertiary Pumps

the topics a, b, and c strike us as logical and orderly because the pumps are being classified on the basis of their importance. But if "d. Centrifugal pumps" is added, it is wrong; a second principle of division is being used. The a, b, c, division uses some such criterion as importance or pumping output, whereas d adds a pointless and confusing criterion—mechanical type.

The mistake of cross-division can be made with any subject and at any level.

If grandmother's pies are classified into quince, mince, apple, open-face, and meringue, is this a cross-division? Of course, because with "open-face" there is a switch from a criterion of content to one of form.

Is there more than one principle of division being used in classifying shoes into leather, wooden, cloth, metal, and paper? No. What is that principle? It is "kind of material used." If health shoes, safety shoes, and children's shoes are added, three more principles of division have been imposed. Now there are overlappings galore, since leather shoes can also be children's shoes, metal shoes are probably safety shoes, and so forth. The classification has lost point and meaning.

Rule 4 applies to every group of coordinates at all levels, from first-order to fourth-, fifth-, sixth-, or any-order headings. The topic-members of *any one set*, although equal in importance, still need to be listed in the way that makes most sense.

Some of the topic-grouping will be done for you by your research work, (p. 18) and by the skeleton outline or format that the company may tell you to follow. Any such aid will probably include the ordering of the first-order headings. Also, whenever you must "write to spec," you will have to prepare your material according to a specific outline (see Chapter V, "Writing to Spec"), and the step-by-step procedures found in specs will automatically help to apply Rule 4. Any ready-made outlines and ordering of topics are naturally of great help to the writer.

The application of Rule 4 for every group of coordinate topics throughout the outline is often aided by simply following through on the *kind* of outline one is making. One's subject and material usually dictate this. Suppose a writer is running a test on smog control or the formation of crystals. This means he is involved in a time sequence; the first part of the phenomenon will probably be reported before the second part and so on. Such an outline, with the report to follow, is of the "chronological" type. The sequence of happening tends to order not only the most important but the least important headings.

Though outlines in technical writing need not involve time sequences, as for example in cases of pure exposition or description of structures, they are usually chronological in nature because they deal notably with reports or instructions. Reports are concerned characteristically with what did happen and thus refer to the past. Instructions deal with what should happen and thus refer to the future. The steps of both are characteristically ordered into time sequences.

Exercise 4. *Order Each Group of Coordinates Properly*

Answer, then see Appendix, p. 167.

1. Is Rule 4 independent of the first three rules? Could these be satisfied without satisfying Rule 4? (p. 26).
2. What aids in applying Rule 4 do we find in technical writing? (p. 27).
3. How does a chronological type of outline help one to apply Rule 4? (p. 27).
4. If an entire outline is chronological in type, will its lower-order headings also tend to be chronological? Need they be? If not, can you explain why they need not be or give such an instance?
5. Why does technical writing so frequently involve the chronological type of outline? (p. 28).

LEARNING THE FOUR RULES

The application of the four rules for organizing one's thoughts is a fair price to pay for producing clear writing. How can we best learn them? They were:

1. Include Every Topic Required by the Subject
2. Exclude Every Topic Not Required by the Subject
3. Working from the Top Down, Divide Each Topic into All Its Subordinates
4. Order Each Group of Coordinates Properly

Let us try to recall the chief considerations that were presented in explaining them. One learns best by using a key idea as a nucleus around which to build a subject.

For Rule 1 we asked, in Exercise 1, what the two main reasons were for failing to include enough in a writing plan (p. 10). Which one of these two was shown in each of the five examples given of violating the rule? In "Machine Shop Theory and Practice" (p. 7), the subheading "Definitions" was found after some headings, but not after similar ones. This shows a lack of parallelism, one of our two main faults. In "Industrial Engineering" (p. 7), the outliner failed to think of an important factor in comparing two ways of buying bulk freon. This shows the other main reason for not including enough in the writing plan, namely, failure to study the subject well enough. Review the other three examples (pp. 7–9) to decide which of the two reasons for infracting Rule 1 each one illustrates.

We asked 10 questions involving Rule 2 in Exercise 2. You may not be able to recall even one of them. So what? There is nothing that prevents you from reviewing, which is the priceless secret of learning. That is all you need to do, spend a little more time building up your processes of recall and recognition. Exercise 2 is on p. 15. Let us review. Out of the 10 questions can you derive *one most helpful thought* for remembering and being impressed with Rule 2? It is probably the three explanations asked for in Question 1 for letting irrelevant topics creep into one's writing plans. It is not only interesting to review these reasons, but they tend to instill in one an active aversion to toting useless, heavy lumber along when writing.

The second most helpful thought about Rule 2 is perhaps the limiting sentence device. As a touchstone for what does and does not belong in an outline, it is too valuable to be disregarded. We should use it, whenever possible, before starting our research.

How about Rule 3, by far the most complicated one? Can you remember any of the 15 questions we asked about it, starting on p. 25?

Let us review them and try to recall our own or the back-of-the-book answers to them. Learning would be much easier if we went about it less gingerly. Do we dislike to review because we are ashamed of not remembering? Nearly everyone forgets with the greatest of ease; to be a good student means, to a large extent, to stop worrying about forgetting and start reviewing.

What one thing will best help us remember Rule 3? Well, what one thing do these two logical processes of subordinating and coordinating agree on and indeed insist on? *Putting each topic (or sentence) in its proper place.* There is a feeling for doing this, a pleasure in doing it well.

The most helpful thought on Rule 4 is perhaps that it applies only to coordinates. After one has produced a set of topics or sentences regarded as equal, one places them in the best possible order. Often, but by no means always, this order is chronological.

TOPIC OUTLINING WITH CARDS

The quickest, surest, easiest way to outline is by using three-by-five-inch cards because, with only a single topic per card, topics can be arranged in any order desired. Here is how to so order them, in seven easy steps.

1. *Topic Sentence*

Write out your topic or limiting sentence (review p. 13), the overall sentence expressing your subject or thesis, on a piece of paper kept before you.

2. *Three-by-Five Cards*

Get a packet of three-by-five-inch cards and sit down at a table. Using your mental resources and the notes gathered from research, write down each idea pertaining to this limiting sentence—let us call it the thesis—on a card of its own, with only one idea per card.

3. *Individual Scrutiny*

Consider each card in turn; if the idea or topic is not clearly comprised under the thesis, throw the card away. (Which of the four rules are you applying here? Rule 2.) Also, if you gather new ideas as you work, evaluate them against the thesis; if they belong, make a card for each. (What rule? 1.) Add and delete cards freely, according to the "all or none" principle; that is, all the ideas that belong in your thesis, but none that do not, should have cards.

Even the most skillful outliner can find that some of the cards are impossible or difficult to classify ("vague designation"). If impossible, he throws the card away. If merely difficult, the card probably belongs somewhere, and rephrasing its topic may show which group it belongs to. The writer should be prepared to rework his topics as much as necessary so as to save the pertinent ones. Sometimes a vague designation can be rewritten into two precise ones.

The most puzzling phenomenon is the topic that seems initially to belong—and often does—to two different groups. If, for example, after card-outlining the subject "Strokes in Operation of a Gasoline Engine," you find a card reading "Compression for the Sake of Power," you have a problem. But since your subject is about various strokes of the engine, the topic must be divided into two topics, namely "Compression" and "Power." Or the topic "Mapping the Earth's Magnetic Field" might better be divided: "The Earth as a Magnet" would introduce this surprising topic to the reader, after which either a *co*ordinate topic on "Mapping the Magnetic Field" (in general) or a *sub*ordinate topic on "Mapping the Earth's Magnetic Field" (specifically) could be made.

4. *Card Classification*

Take up the cards again. By throwing one after another on the table, separate the cards into stacks or groups, according to what seem to be the main divisions (first-order headings) of the thesis. Then place the card that names each group at the top of it so you can easily see your first-order headings. Do not worry about arranging the lower-order headings yet. Concentrate on forming one stack of cards for each presumed first-order heading and on seeing that each card gets into its proper stack.

In effecting Step 4, you may find you do not have a satisfactory set of first-order headings. This can be discouraging, since all research is presumably completed, and only the organizing of your topics remains. Fortunately, whenever pertinent subtopics lack a heading, the mind can supply it. (Doing so expresses Rule 1.)

For example, suppose you were ready to write about crop-spraying, but the first-order headings you had collected did not represent the subject. You simply remedy the defect by "straight thinking." The idea of dividing your subject *geographically,* for example, could make you think of the understandably different crop-spraying situations in the East, South, Midwest, and West—four first-order headings!

Then, to think of crop-spraying in terms of time would let you discuss it from the points of view of past, present, and future. Three more good ideas. Four geographical, times three chronological, gives you twelve new headings!

5. *Organization by Stack*

Spread out each stack vertically. By comparing the cards in each stack with one another, make sure they belong together. What you should now have at the top is your set of first-order headings, one for each stack; below them—mixed up, of course—will be all the lower-order headings that belong to that stack. Taking one stack at a time, apply the principle of looking for the *next*-order headings in each. These would be those that each first-order heading divides into—its second-order headings. Separate them vertically into their own substacks, then place under each substack the topics that belong to *it*.

In outlining, you continue the dividing process but you keep changing the topic divided, in working from top to bottom. You change the location of various cards at will, as you make up or change your mind as to what topic belongs where. Whenever a substack or a stack is in final order, a paper clip can hold its constituent cards together.

In thus organizing a stack of cards, you have placed subordinated cards behind their superordinates. You have applied Rule 3 thoroughly.

6. *Putting the Stacks in Order*

Now you decide on the order of the main stacks themselves and rearrange accordingly (Rule 4). What topic should be the first of the first-order headings? The stack with this topic heading may be placed at the far left on the table, the stack headed by what you decide is the *second* of your *first-order* headings, just to the right of the first stack, and so forth.

7. *Surveying the Outline*

Now show yourself the entire finished outline by copying down your ordered topics on a sheet of paper. Since your cards have served their purpose, you may throw them away. Carefully study the outline. You did not see it before as a unit, and it may be imperfect. If so, change it while there is still time.

In Rule 3, we discussed subordinating and coordinating processes by themselves, not in connection with any card-outlining activity, since the latter is simply a *means* of outlining. Using the cards, in other words, does not free us from making a single subordinating or coordinating decision, but merely enables us to make these decisions and to change our minds about them more easily.

Steps 1, 2, and 3 of "Topic Outlining with Cards" (p. 30) show how we collect the topics to be arranged. The internal organizing of each stack begins with Step 4, which is the first major act of subordinating. Step 4 includes the *thinking up* of first-order headings to fill a lack of such headings. In Step 5 the internal organizing is finished and every card in the stack has its proper place. In Step 6 the stacks are placed in a desired order so that what should be the first of the first-order headings is really first, and so on. In Step 7 the card outline is transferred to a sheet of paper so that one can survey his handiwork and write from it easily.

CLASSIFYING DEVICES OF THE OUTLINE

All final decisions on subordinating or coordinating are reflected in the appearance of an outline. Otherwise it would be neither a record of these decisions nor a directive showing what and how we have decided to write. Three ingenious yet simple practices provide this needed record, giving the outline its characteristic look. These are indentation, numbering, and lettering.

Indentation indicates a heading of less importance, at all levels. Thus in

> III. Experimental Apparatus
>   A. Procedure

and in

>   A. Procedure
>   1. Preparation and Loading of Cell

the subordinations of A and 1 are shown by a slight placement to the right. (There would be a III.B and a III.A.2, of course.)

Starting two or more same-order headings at exactly the same distance from the lefthand margin, that is, indenting similarly, indicates the logical process of *co*ordinating, as for example,

> III. Experimental Apparatus
> IV. Container

or

>   1. Preparation and Loading of Cell
>   2. Final Inspection

Both numbering (I, II, III, and 1, 2, 3, etc.) and lettering (A, B, C, and a, b, c, etc.) constitute a *second* means of subordinating and coordinating. By arbitrary convention, large Roman numerals precede more important headings than do capital letters; the latter precede more important headings than do Arabic numerals. In addition, *contrasting* one type of numbering or lettering with another type (as I with 1 or A with a), or contrasting a type of numbering with a type of lettering (as 1 with a) can indicate the latter-type topics as subordinate.

Numbering or lettering in the same way indicates that the topics so designated are of equal importance.

The generally accepted method of using numbers and letters in outlining is as follows. First in importance is the large Roman numeral—I, II, and so forth. Second-order headings are shown by capital letters, A, B, and so forth.

The following example continues this method, showing the profile that a symmetrical outline—only two headings in each order—would present. Such symmetry is unusual.

    I. First-order Heading
       A. Second-order Heading
          1. Third-order Heading
             a. Fourth-order Heading
                (1) Fifth-order Heading
                    (a) Sixth-order Heading
                    (b) Sixth-order Heading
                (2) Fifth-order Heading
             b. Fourth-order Heading
          2. Third-order Heading
       B. Second-order Heading
    II. First-order Heading

There can be almost as many different outline-profiles as there are outlined writings. We say "almost as many" instead of "as many" only because two outlines, although about different subjects, might *happen* to have the same outline-profiles. For example, each outline could contain only four first-order headings, two second-order headings under the first first-order heading, and nothing else. But their similarity of profile need not, of course, indicate any similarity of content.

So the way a group of headings looks is merely a reflection of the analysis and ordering-judgments already made by the writer. An outline mirrors a set of logical decisions. It is a value scheme; in thinking it out, the writer assigns headings so that they signal for him the relative importance of the associated topics. Thus any of the orders of numerals and letters might occur from two to a great many times in an outline, depending on what the subject dictates. The subject of automobiles, for example, might be divided for treatment into those powered by steam, electricity, gasoline, and atoms. The steam heading might contain two subheads, while the gasoline heading might contain many more.

The various types of headings, degrees of indentation, and other devices of the outline usually appear in the body of a *text* quite differently from the way they appear in a table of contents. This is especially true of type sizes. Please note this by examining a few books at random.

Notice that capitalizing the letters of headings is a further means of indicating importance. In the written text itself, underlining is often used to make certain orders of headings stand out.

Seventh-order headings are rare; they can be the small Roman numerals i, ii, iii, iv, v, and so on. Fifth- and sixth-order headings are also rare.

The routine use of all these designations should be easy. You might write the list a few times, for drill:

I, A, 1, a, (1), (a), i

To briefly review, we have noted that the topic or statement following I in an outline is called the first of the first-order headings. There should be at least one more first-order heading in the outline (p. 16). Is there in the preceding outline (p. 34)? There is, and it appears correctly on the page at an equal distance in from the left-hand margin.

There should be more than one second-order heading in the outline; and they should all appear at the same distance from the left-hand margin as does the first. At least two members of each type or level of heading are placed where needed throughout an outline.

Will the "profile" or topography of an outline always show the same symmetry that our example does? Probably not, because the number of headings of any type in an outline depends on considerations of writing content.

Exercise 5. *Profiling an Outline*

An outline might have only two first-order headings, the second of which has no subdivisions (i.e., second-order headings). But the first of the second-order headings might have four third-order headings. The second of these third-order headings might have three fourth-order headings. The profile of the outline would not be symmetrical. How would it look? Please write up, on scrap paper, the way such an outline would look; then compare your version with the Exercise 5 answers, p. 168.

What would be the first of the first-order headings in an outline? It would be whatever heading or topic *followed* I. The designation or symbol, such as A or 1, and so on, is not the heading itself.

How would you *designate* the second of the third-order headings? Since third-order headings are indicated by Arabic numerals, the second such designation would be 2.

How is the eighth fourth-order indicated? By a small h.

Exercise 6. *Correct Outline Symbols*

What are the correct symbols for the following?

1. The seventh of the fourth-order headings.
2. The third of the second-order headings.
3. The fourth of the seventh-order headings.
4. The first of the second-order headings.

5. The second of the third-order headings.
6. The third of the sixth-order headings.
7. The last of the fourth-order headings (trick question, but please answer).
8. The only third-order heading (trick question).

Write out your answers on scratch paper, then check them in the Appendix, p. 168.

We have been talking about the classic "number-letter" method of outlining. Another leading practice, the "decimal" or "spec" (for specification), might better be termed the "dot" method. Two such dot examples appear in the Appendix, pp. 169, 171, and should be carefully studied by the reader. They are pertinent to Chapter V, "Writing to Spec."

THE SENTENCE OUTLINE: ANALYSIS AND COMPARISON

What you have prepared in the preceding seven steps (pp. 30–32) is a topic outline on a single page. Each topic resembles the subject of a sentence, the rest of which, if expressed, would constitute its predicate. Without a predicate, no statement about a subject can be made. If you take the trouble when outlining to add predicates and other phrases to your topics, thus forming a sentence outline, you will have thought out the main statements you are going to make when you write! With just a topic outline to guide you, you will have thought out what you are going to write *about*, but not what you are going to *say*.

To appreciate this interesting situation more fully, let us compare and contrast part of a *topic* outline with the corresponding *sentence* outline of the same subject, "The Lost Island of Atlantis." Diagonal lines indicate the entries that correspond.

Topical outline

I. RENEWED INTEREST IN ATLANTIS
  A. SUPPOSED LOCATION
  B. SIGNIFICANCE OF ATLANTIS

Sentence outline

I. Today a renewed effort is being made to discover the remains of Atlantis.
  A. It is thought to have existed somewhere between Spain and America.
  B. Scientists have felt that if its existence could be proved many mysteries of American and European race similarities might be solved.

II. ANCIENT INDICATIONS —— II. The existence of Atlantis is
    OF ITS EXISTENCE   strongly indicated by evidence
    A. WRITERS   from the ancients.
      1. Plato   A. The writers describe it in
      2. Himiles   connection with the Sargasso
    B. MAP-MAKERS   Sea.
      1. Boccario   1. Plato mentions Atlantis
      2. An unknown map-   in two works.
         maker of the 15th   2. Himiles said the Sargasso
         century   Sea had no depth because
                           the remains of Atlantis
                           were just beneath the
                           surface.
                      B. Map-makers of old portrayed
                      Atlantis.
                      1. In a map by Boccario
                       dated 1435, Antilla (At-
                       lantis) is clearly marked.
                      2. In another 15th-century
                       map a large group of
                       islands, corresponding to
                       the West Indies but much
                       larger, is shown.

III. RACE AND LANGUAGE ——III. Peculiarities of race and lan-
    INDICATIONS OF ITS   guage on both sides of the At-
    EXISTENCE   lantic Ocean seem to support the
    A. THE AMERICAN   possible existence of Atlantis.
       INDIAN ——————————A. The variety of coloring in the
    B. THE BASQUE   American Indian may have
       LANGUAGE   been caused by the fact that
                          Atlantis was said to have
                          been inhabited by red, yel-
                          low, white, and black races.
                      B. The Basque language is un-
                       like any other European
                       language.

Notice how much more information the sentence outline gives. Contrast the topic "I.A. Supposed Location" with the sentence "I.A. It is thought to have existed somewhere between Spain and America." Or contrast the left-hand I.B., II.B.2, and III.A with their sentence counterparts on the right-hand side. What a difference in information and interest between the two outlines! You would not know from the topic outline, for example, that the lost island of Atlantis is supposed to be in the Sargasso Sea—certainly an intriguing bit of information.

Is the topical outline, confined as it is to never making statements, really as limited as this? Well, of course a topic can be beefed up with the phrases and clauses adapted from its counterpart sentence. Instead of "Supposed Location," we could say "Supposed Location Somewhere between Spain and America." But including all such information in a mere noun phrase can be cumbersome. II.A.2, for example, would have to read "Himiles' Statement that the Sargasso Sea Had No Depth Because the Surface of Atlantis Was Barely Covered with Water." What an unwieldy topic for a report! To go to such lengths to avoid making a sentence outline would be ridiculous.

Compared to the sentence outline, the topic outline is limited.

Changing a Topic into a Sentence

How much work is needed to convert a topical outline into a sentence outline? Remember this fragment?

I. Strokes in Operation of a Gasoline Engine
   A. Intake
      1. Volume Increase
      2. Pressure Decrease
      3. Entrance into Cylinder of Gas Vapor and Air
      4. Closing of Intake Valve

The words necessary to change the foregoing into sentences are in italics, as follows. The words not needed because of being rephrased are *cancelled by dashes.*

I. *There are four* Strokes in *the* Operation of a Gasoline Engine.
   A. *At the* Intake *Stroke, the Piston Moves Down.*
      1. *The* Volume Increase*s in the Space Above the Piston.*
      2. *The* Pressure Decrease*s in the Space Above the Piston.*
      3. ~~Entrance into Cylinder~~ of Gas Vapor and Air *Enter the Cylinder.*
      4. ~~Closing of~~ The Intake Valve *Closes at the End of the Stroke.*

Here we note that the changes needed to convert a topical outline into a sentence outline are various. One does not always simply add a predicate. Note also that, although some of the changes are purely verbal and add nothing to our knowledge, others do. To say "Gas Vapor and Air Enter the Cylinder" is almost like saying "Entrance into Cylinder of Gas Vapor and Air," it is true. But to say "At the Intake Stroke, the Piston Moves Down," tells us much more than just "Intake."

Making sentence outlines is worth the extra trouble. Far from adding work, the sentence outline decreases it.

Advantages of a Sentence Outline

1. Outlining by making key statements helps one grasp his subject, thus clarifying thought processes, minimizing vagueness, and making the final writing easier and faster.
2. If one who is writing for the immediate approval of a superior shows him a sentence outline, he gets the clearest possible picture of what is intended; this may save much rewriting.
3. The sentences in a sentence outline are the logical topic sentences for paragraphs or key statements for entire strings of paragraphs.

All in all, sentence outlines are preferable in general to topic outlines. But sentence outlines are inadvisable when pro-and-con discussions of topics are desired or when much information or many instructions are given, as in manuals. In both these cases the proving of certain points of view is not needed, but rather the comprehensive presenting of subjects. Noun phrases serve best for outlines in both these cases.

But the sentence outline should be built from a topic outline, not only because the topic is the natural subject of the sentence, but because the card method lends itself to topic-forming rather than to sentence-making.

AN UNEXPECTED DIVIDEND

The stock you have taken in outlining may surprise you now by declaring a dividend. Here it is: the best way to *review* a subject is to sentence-outline it, using the same three-by-five-inch cards. Doubtless it was by working from an outline that a textbook author wrote whatever it is we may wish to review; as students, we reverse the process by outlining for ourselves what he has written.

To learn an assignment we have just read we want to know, above all, the main thought. This should be the author's topic sentence. But we may *end* with it. True, writing from an outline and learning by outlining use the same process, but in reverse. That is, the author started with an outline, and ended with his written book. His reader starts with the book and builds a mental outline of its contents.

In using our outlining skills for the purpose of learning, we start on the first page to be reviewed and write down the most important statements, *one per card,* just as we did before. We work in this way from the first page of the assignment to the last. The forming of every sentence is part of the learning as we keep deciding what is important, and it is sentences we decide about and write down—a mere topic *says* nothing!

Notice that the advantage of using separate cards for outlining still pertains. We start reviewing with the first page, using a separate card for every key statement, and end with a collection of cards containing the key thoughts. By this time we have already half learned the lesson.

Proceeding as before, we group our cards so as to subordinate the minor thoughts to the major ones. Now we are in the very lists of learning, where competent learners are separated from poor ones. The former finish reviewing a lesson with the key thoughts firmly in mind; the latter fail to follow through, to subordinate the minor thoughts, and thus remember them as on a par with the major ones. Their picture of the subject is accordingly a worm's-eye rather than bird's-eye view.

As we so well know by now, all subordinating and coordinating judgments take effort. What we are doing here is winnowing, separating grain from chaff. But in thinking by this reverse outlining trick, it is as though we were using the latest mechanical agricultural device instead of the rude flail of our forefathers. We are minimizing our effort, maximizing our return.

Although we arrange the cards according to *our* purpose and conception of the subject, we usually find that we have distilled from the text the best that the author had to give us.

Exercise 7. *Topic and Sentence Outlines*

As usual, first answer from memory, then look up text reference, then turn to answers in Appendix, p. 170.

1. State the difference in structure between a topic and a sentence outline (p. 36).

2. How is it possible to outline what one is going to write about, but not what he is going to say? (p. 36).

3. Does the topic outline tell you where Atlantis was believed to be? Does the sentence outline tell you? How do you explain the difference? (p. 37).

4. What things must be done to change a topic outline into a sentence outline? (p. 37).

5. Can you estimate about how much longer it would take to prepare a sentence outline than the topic outline from which it was made?

6. Instead of thinking up statements when outlining, do we content ourselves with just topics because we do not yet know what we want to say about these topics? If so, should we not know this at the outline stage? Why might it pay to take the extra time? Because we should do our important logical (and not merely compositional) thinking before rather than during our writing? (p. 36).

7. Mention three reasons to outline by using sentences rather than topics (p. 39).

8. When would it be preferable to prepare a topic rather than a sentence outline? Why? Do you believe your answer explains why there are so many more topic outlines than sentence outlines? (p. 39).

9. In regard to procedure, what is the difference between a writing outline and a learning outline? (p. 39).

10. In terms of successfully making learning outlines, how would you express the difference between efficient and inefficient learners? (p. 40).

CHECKLIST TO IMPROVE OUTLINES

The following questions will help to perfect an outline. Imagine that you have written an outline, and after reviewing it, must answer "no" to each question. Then answer the question, "Why would this fault tend to make a poor outline?" After that read the text immediately following the word "no." Let us illustrate what we mean with the first question.

1. *Does the Entire Outline Express a Definite Aim?*

"No." Then why would this fault tend to make a poor outline? The answer is that the aim or purpose of the finished product would be unclear to your readers. If the aim is not clear to you, the writer, or if you have not expressed it in the outline, you cannot expect your readers to see it.

What can you do to correct this fault while your thinking is still in the outline stage? You can apply your knowledge and use the devices already discussed in the text. Did you use a limiting sentence designed to express a definite aim? Remember that a sentence outline will force a writer to develop his thinking and his conclusions while still planning what to say.

2. *Does Your Outline Give the Essentials of the Subject?*

"No." This would simply mean the outline was superficial and failed to *qualify* as an outline of your subject.

3. *Does the Outline Cover the Subject?*

"No." If not, it is not complete, comprehensive. An outline should represent *all* of a (carefully delimited) subject.

4. *Is the Outline Clear Throughout? Do the Headings Make Sense?*

"No." Then you should not write from such an outline; if you do, the writing will not communicate.

5. *Has Each Heading Been Sufficiently Developed?*

"No." If not, the outline will appear sketchy. You should not be afraid to say the obvious; this can always be done in a special, interesting way.

6. *Are the Main Headings the Most Appropriate Ones for Your Purpose?*

"No." If not, the purpose will suffer.

7. *Are All Useless Headings Avoided?*

"No." If not, emphasis will be lost and the reader's interest will flag. Maybe some of the topics should be played down. Remember that one must be ruthless here, regardless of research done and time spent.

8. *Do the Groups of Headings Indicate How Long, Relatively, the Parts of the Finished Work Will Be?*

"No." If not, the writing may be seriously disproportioned. If you have fully worked out certain parts of your outline, perhaps down to third- and fourth-order headings, one would expect the consequent writing to reflect that attention. If not, there would be a disproportion between that part of the outline and its written result. Obversely, if a skimpy outline was followed by a surprising amount of writing, one would suspect that not all that had been written had appeared in its outline.

9. *Does the Outline Give a Sense of Continuity and Unity Instead of Being a Mere Collection of Headings?*

"No." This can be remedied by thinking stubbornly of your reasons for juxtaposing headings; by thinking, before beginning to write, of transitional

thoughts between headings; and, by making sure that the order chosen
is the best possible.

10. *Does Each Heading, if Divided, Have at Least Two Subordinate
Headings?*

"No." If not, an impression of careless thinking will be given. It will seem
to the reader that either the part really belongs to the heading just preceding
(its superordinate) or that the part should have a coordinate.
   This is, for example, incorrect:

> B. Mass-Energy Equivalence
>   1. Einstein's Unified Field Theory

For if B has but one part, that part seems identical with B. If B has
two or more parts, these should be stated. For example:

> B. Mass-Energy Equivalence
>   1. Work in Nuclear Changes Before Einstein
>   2. The Unified Field Theory

11. *Are There Less Than Six First-order Headings?*

"No." If not, some may be subordinate to others, but mistakenly grouped
as coordinate because of a failure to finish the difficult analytical thinking
involved. That is, an outliner sometimes has a tendency to let up; he
knows certain topics belong to the outline, but not exactly where. So to
save work he coordinates them with others, whereas some should be subordinated. ("Six" is not a magic number; it is an arbitrary point at which
to start wondering if there are too many coordinate headings.)

12. *Does Every (Subordinate) Heading Belong Under Its
Superordinate?*

"No." If not, it belongs under another heading or perhaps does not belong
at all.

13. *Is Every Group of Coordinates Free of "Cross-division"?*

"No." Then clarity is threatened. Cross-division is the use of more than
one criterion in choosing the subordinates of a heading. For example, if
in reading about the five methods of printing—relief (letterpress), planog-

raphy (usually photo-offset lithography), intaglio, stencil, and photography—we ran into a discussion of type composition, printing, and binding, we would be confused. The writer has begun using a criterion other than methods of printing for his subject, namely, a chronology of processes that applies to several but not all five printing methods. Even if it applied to all, he would be wrong because cross-division renders a subject unclear.

How many criteria of division are being used in misclassifying suits into double-breasted, single-breasted, herringbone, imported, two-pants, and lightweight? There are five.

Moral: use only one criterion in analyzing a heading into its constituent parts.

14. *Are Parallel Treatment and Parallel Wording Used Whenever Advisable?*

"No." If not, equality of ranking and coordinateness of meaning are made more difficult to grasp and a carry-over of interest to new topics is impeded. The equality of coordinate headings should be confirmed by expressing them grammatically in parallel ways.

For example, if under I. B. Compression (p. 8) we say "1. Volume Decrease," we should not place under it "2. Increase of Temperature," but "2. Temperature Increase." Differences in the wordings that should indicate similar facts is confusing to one who is organizing his thinking. And these differences so easily creep in. To illustrate:

    I. How Heat is Produced
   II. Measurement of Heat
 III. Heat Transmission

These coordinate headings should have parallel wording:

    I. Production of Heat
   II. Measurement of Heat
 III. Transmission of Heat

    I. Heat Production
   II. Heat Measurement
 III. Heat Transmission

    I. How Heat is Produced
   II. How Heat is Measured
 III. How Heat is Transmitted

Subjects to be outlined for reports, instructions, and so forth, vary considerably as to the ease with which their constituent topics are classified. Some subjects can be analyzed rapidly into major and minor headings, properly connected; others will not come apart easily. Whereas one outline can be a triumph of logical analysis another, despite stubborn effort, may resemble a mere listing.

The outline has great flexibility. It can present topics in a variety of ways. Sometimes, as in historical treatment or other narrative, the topics follow an easy chronological order. An outline may also provide for continued exposition or description of highly complicated machines and processes. It can structure a strictly argumentative report from basic considerations, premises, and reasons to final conclusions.

Concentrating on different *parts* of the subject in a single outline can strengthen rather than weaken it. For example, an outline about a certain rocket engine (and the resultant report) might begin by *analyzing* its operation, then switch to *illustrating* its successful flights, and end by *arguing* for its superiority over competing models.

The value of working up outlines reminds one of "a stitch in time saves nine." Outlining saves rethinking and recasting later; if an author rewrites, it is because he must, either as a directed employee or as a slave to his conscience. The point is that outlining is well worth the trouble; it achieves a superior writing product and minimizes rewriting. Or, to mix sewing with hiking, what may seem the longest way 'round is always the shortest way home.

A completed outline is like a power tool in the sense that it puts stored-up energy—that of having made all those logical decisions—at one's fingertips. When he begins to type, the writer converts this organizing work into words; and he half knows what he wants to say (especially from a sentence-outline) because he has thought about it before, as he made his outline.

The reader has by now given outlining much time and effort. But his investment should pay off heavily. Outlining tends to become a pleasure and to make a strong writer out of a weak one.

SUGGESTED FINAL FOR CHAPTER I

After working out your answers, the pages indicated may be turned to. At this point, we pose twenty instead of ten questions.

1. Why should anyone planning a technical or scientific career make sure that he learns how to write well? (p. 1).

2. Explain why technical writing is different from the usual college theme-writing. (pp. 2–4).

3. Why is it more difficult in the long run to write without planning what you are going to say? (pp. 5, 6).

4. Do you think it likely that a faulty writing plan will somehow blossom into a good piece of writing? Why is it wise or unwise to try to carry a writing plan in your head? (p. 6).

5. Explanations have been offered as to why irrelevant topics .and data creep into our materials for writing. Have you noticed that this happens to you too? If so, describe how it seems to occur in your case.

6. State the main facts on the use of the limiting sentence in outlining. Which of the rules for outlining becomes much easier to apply with the aid of a limiting sentence? (p. 13).

7. "A topic never has more than one subordinate." Is this statement true or false? Why? (p. 16).

8. Why is part of our outlining work, namely, the ordering of some of the topics, apt to be done for us before we begin formally to outline? (p. 18).

9. Why was infraction of Rule 3 in outlining said to be more serious than infractions of the other three rules? (p. 18).

10. "If one has just noted that Topic A is coordinate with B, he should at once see if A is also coordinate with C, D, and so forth. Likewise, after discovering one subordinate, he should seek for others." Are these statements true? Why or why not? (p. 16).

11. Can you explain how outlining work can be generally the same for everyone, yet individually varied? (pp. 16, 21).

12. What is an outline, and why is outlining worthwhile? (p. 24).

13. Why does the chronological type of outline predominate in technical writing? (p. 28).

14. Write the outline skeleton that has five first-order headings; three second-order headings; eight fourth-order headings, three of them being under the second of the third-order headings; four third-order headings; and two fifth-order headings.

15. What are the correct symbols for the following?

1. The sixth of the third-order headings
2. The first of the fourth-order headings
3. The last of the second-order headings
4. The third of the fifth-order headings
5. The tenth of the first-order headings

16. Is a table of contents (for a book) an outline? (p. 24).

17. Make a few general statements as to what things must be done to change a topic outline into a sentence outline. Can you imagine *not* being able to use the statements in a sentence outline in the writing itself? (p. 39).

18. Construct a brief sentence outline for any subject with which you are familiar.

19. What is cross-division of topics? How does it relate to clarity in outlining? (p. 26).

20. What is the importance of parallel grammatical treatment in outlining and when should it be used? (p. 44).

II

How to Write Effective Sentences

One can rapidly learn to *write* (after merely planning) by deliberately correcting, several times, *the few basic mistakes* that poor writers make. Naturally one can then avoid these mistakes when doing his own work and can thus build into himself, for himself, the sound practices used by professionals. To accomplish this, each of only seven basic rules will be taught on the following pages by a "5-5-10 system": five errors are corrected as one reads, five more by means of student-plus-Appendix, and the final 10 (or more) by the same method. The conscientious student can soon make of his writing a great new skill and pleasure. It takes work, of course, but not much considering the gain. Care to try?

CURING SICK SENTENCES—SEVEN SPECIFICS

Following is the method to be used in presenting the various rules of sentence structure. We shall give, analyze, and correct five examples of the first sentence fault, the use of unnecessary words. Five more examples will follow, to be corrected on scratch paper. You are then to compare your corrected versions with those in the Appendix. In the next step 10 final examples of sentences of unnecessary wordage are presented. You are to correct these, also on separate paper, before turning to the Appendix.

Rule 1. *Use Only Necessary Words*

Can one write a textbook which will be fun to read?
Can ~~one write~~ a textbook ~~which will~~ be fun to read?
Can a textbook be fun to read?

49

Very few persons write as briefly and simply as they should. In the foregoing case, one need say nothing except, "Can a textbook be fun to read?" Obviously it must be written by someone; therefore "one" should be taken for granted.

Yet we are prone to overwrite, and there are many ways to do it. For example, if the sentence were "Can a textbook be written that would be fun to read?", we would delete "written that would be" and wind up with the same ideal sentence, "Can a textbook be fun to read?" Or if we had "Can there be a textbook that is fun to read?", we would delete "there be," change "that is" to "be" and have the same sentence again.

As the other rules show, improving a sentence does not always require that words be dropped. They may need to be changed or added to. Sometimes a writer will insist that he needs certain words to round out or emphasize his thought. When this is true, there is no problem; let the fortifying words be used. But usually this insistence is just an excuse for lazy writing.

Also, students with literary ability have a problem here. Literary values are not confined to what must be said; they are psychological luxuries, dividends, new values given to the reader by the writer. To limit literature to what is necessary would send us all to the cultural poorhouse. In technical writing, however, we set down only what is necessary.

Thus, at first, the more literary minded find it difficult and distasteful to avoid using "extras" in their writing. But they must learn to do it. A praiseworthy style in technical writing consists of the simplicity and strength resulting from taking these rules, especially 1, seriously.

To ignore Rule 1, to permit nonpaying passengers, deadhead words, to ride along in our sentences blurs their outline and thus spoils comprehension and destroys the emphasis otherwise securable. This failing can be combatted in two ways: After writing, we can cross out what on second thought is unwanted in a sentence, or we can think out our original thought so carefully that we finally write down only what is needed. The latter trick is developed only after much practice. So we shall correct the overwritten sentence. This will help us form the habit of thinking only working words in the first place.

Here is our first official group of five words, analyzed in the text.

We do not doubt but that the project will succeed. The word "but" is not needed. We should say "We do not doubt that the project will succeed." All the words left are required by the meaning, as can be seen by trying to remove one. The only other word (besides "but") we could come close to leaving out is "that," which is not good. We would have, "We do not doubt the project will succeed." The meaning is fairly clear, but the sentence does not sound finished enough.

This type of sentence containing an unnecessary word indicates poor usage. Here are a few examples.

1. Take your lab coat off of the hook and start to work.
2. The solution will evaporate faster if you cover it over.
3. You may not leave the classroom any time you want to.
4. A good technical man rarely ever likes to teach.

Delete "of," "over," "to," and "ever" from the sentences and they will sound trimmer, more professional.

The six motors that were destined for the main plant were shipped yesterday. We do not need "that were"; they make the sentence less effective. Read it both ways and you will probably agree that the shorter is the better one. There is, however, no virtue in mere contraction; it must improve the sentence. (Sometimes more words are needed, rather than less—but this involves Rule 2.) The sentence as corrected is simply crisper, more emphatic. One reason for this being so is that one "were" was dropped. Repeating the sound of "were" was slightly unpleasant and slowed up the sentence, which we will call the "correct but less effective" type.

He is studying the field of biology. Here we have a slightly different situation. "The field of" *seems* necessary; it takes a little thinking about such expressions to realize that, although harmless, they are also pointless. "He is studying biology" is all that one needs; everybody knows the subject of biology is a field. If the writer wanted to stress that the student was studying the entire field rather than a specialty, he should have managed to say it, perhaps with "He is studying general biology." The type may be termed "seemingly necessary."

This number is suggested as being the best estimate of the value. Here we come to the type of "unnecessaries" in which correction is debatable. Suppose we leave out "suggested as being," in the belief that the writer means simply that "this number *is* the best estimate of the value." He may say he does not mean that and wants merely to *suggest*. But it is surprising how often the writer, on second thought, will agree that he prefers the simpler, more direct version and that to spin the sentence out, adding conditions, expressing diffidence, and so on, is just poor writing. If he insists that "suggested as being" expresses a meaning he wants in the sentence, then the analysis changes a bit; it becomes necessary to see if there is a choice of words better than those used. The writer may mean "probably," in which case he should say it. We labeled this the "debatable" type.

The building is used for residential purposes. To free this type of sentence from unnecessary words results usually in shortening it, but it requires rewriting. First one habituates himself to spot and then question the need for inflated phrases like "for residential purposes." Then he must be willing to locate, in the language, a simpler way of saying this. He remembers that the tiny word "as" can convey the idea of purpose, so he tries "The building is used as a residence," and likes it better. He has shortened the sentence, but only by rewriting it. He lets it sound in his mind a few times and realizes each time that the more pompous way of saying it detracts from the neatness and strength of the statement. These thoughts and rewritings become easier with practice.

Now for a second group of five sentences about our first rule, "Use only necessary words." Please correct them on scratch paper. Delete unnecessary words, and otherwise alter; write out each amended sentence and compare it carefully with the original. Then compare your final versions with those in the Appendix, p. 172. As the student corrects these next five sentences, he can increase his mastery of Rule 1 by seeing if they too are examples of the five types already noted, namely, "poor usage," "correct but less effective," "seemingly necessary," "debatable," and "requires rewriting." Some of these types overlap, but thinking about them helps us remember the rule.

Exercise 8. *Use Only Necessary Words*

1. During the course of the experiment, he resigned; but he was asked to stay on until it was completed. (Type?)

2. A test installation was constructed so as to permit operational check-out of the handling equipment. (Type?)

3. Since I know the check will bounce anyway, the customary practice of endorsing it on the back is entirely unnecessary. (Type?)

4. He continued to remain at his old post. (Type?)

5. Engage foot lever to compress crimping dies; then, when foot lever is down, the crimping dies will be in a closed position. (Type?)

We hope that working with these sentences (note what an odd and interesting language English is—"working with these sentences," all of it, is being used as the subject of a clause still unfinished!) has strengthened the habit of analysis. Writing, unlike talking, is an *unnatural skill,* acquired with difficulty. These little exercises, sprinkled throughout the text, should help to make of writing a kind of arduous pleasure instead of a bugaboo never faced correctly and mastered.

The following is a longer, formal test on the same rule, "Use only necessary words." Write your corrections on another sheet, then turn to the Appendix, p. 173, for the answers.

Exercise 9. *Final for Use Only Necessary Words*

1. Good weather conditions prevailed throughout the flight.
2. Hurry and connect up the battery.
3. We encountered frustrations in applying the plan, and experienced many difficulties.
4. The next installment will deal with radioactivity and its relation to nuclear energy—in a forthcoming issue.
5. Final completion of the run is scheduled for February.
6. The main coolant system is utilized to conduct the reactor heat.
7. He completely ignored the fact that we had not received a work order.
8. By this procedure it was established that defects could be removed.
9. The purpose of this report is to summarize the costs incurred for engineering, construction, and testing.
10. A thorough chemical analysis was performed on chips taken from the control ingot. This analysis appears in Table 4.

Rule 2. *Avoid Incomplete Constructions*

One may fail to express his desired meaning by leaving out part of a sentence. Though the meaning of the sentence can often still be grasped, irritation that remains may impair the reader's interest. In the worst cases, the sentence cannot be understood.

Here are five examples of incomplete sentences.

He was always studying strange maps, like Asia. The first reaction to such a sentence, since the meaning can be extracted easily, is just a vague feeling of something being wrong. But an accumulation of such vague feelings while reading a long report can ruin its chance for a strong impression and defeat its purpose. One avoids composing such sentences by being very literal. Does one want to say that "he" was studying Asia itself, and that Asia itself is like a strange map? No; one means, of course, "He was always studying strange maps, like that of Asia." More words ("that of"), but needed.

The use of an organic fluid in a nuclear reactor is in certain respects preferable to sodium. This is an omission of the same type, but much harder to spot. One is comparing the use of an organic fluid with the element sodium itself, when the purpose is to compare two uses. So again we insert "that of" between "to" and "sodium," and have a completed sentence. We should never economize on words when the clarity of what we are saying is thereby impaired.

He dislikes supervisors as much as his colleague. Here a stinginess with words blacks out the meaning of the sentence. Does he dislike supervisors as much as he does his colleague, or does he dislike supervisors as much as his colleague does? There is no way of knowing until the writer decides which meaning he intends, and then adds either "he does" between "as" and "his" for the first case or tacks on "does" after "colleague" for the second case. If the writer *happened* to mean both, there would be no way to avoid saying, "He dislikes supervisors as much as he does his colleague and as much as his colleague does."

Looking for a suitable location, the truck broke down. We usually know what is meant by such statements, but they make us smile. Here we imagine a neurotic, unhappy, and no doubt abused truck seeking a good place to have a breakdown. We know that "looking for a suitable location" should refer to something in the sentence. The only "thing" it could refer to is "truck," hence the humor.

"Looking" is a dangling participle. The correct term for this general type of misstatement is "dangling modifier." Infinitives (*"to work* as a mechanic, skill is necessary"—skill does not do the work!), gerunds ("after *seeing* the doctor, his stomach stopped aching"—stomachs cannot see!), and even clauses ("while visiting in Cucamonga, a cloudburst occurred"— these nimbus clouds get around!) can also dangle.

We classify dangling modifiers under Rule 2 because they are incomplete constructions. Notice how, in each case, the trouble is caused by leaving out needed words. The action of looking for a suitable location, for example, should be performed by someone. Up to now, only a truck is available. To correct this dangling modifier, we merely add a subject that can perform the action: "While *we* (or "they," etc.) were looking for a suitable location, the truck broke down."

While asleep at the lecture, a thief picked my pocket. This dangling modifier "while asleep at the lecture" is one of the more glaring examples, since we see at once that it dangles. The thief was scarcely a somnambulist. To correct, we simply apply the rule of supplying the noun or pronoun that being asleep refers to. The intended meaning tells us what this is. The "my" of "my pocket" tells us that the first person singular pronoun will be correct. So we say "While I was asleep at the lecture, a thief picked my pocket."

Usually there are alternative ways to correct for dangling modifiers. We could say, "While asleep at the lecture, I had my pocket picked by a thief." In this case, we made a different independent clause, with "I" as the subject, instead of "a thief." In either case, the dangler has been removed.

Dangling modifiers, caused by the writer skimping on words, end up designating inanimate objects rather than human beings as performers of actions. In our three previous examples, "skill," a "stomach," and a "cloudburst" were designated as agents for working, seeing the doctor, and visiting Cucamonga. To correct these three, we reintroduced the humans who had been hustled away: "To work as a mechanic, *one* requires skill," "after *he* saw the doctor, his stomach stopped aching," and "while *we* were visiting Cucamonga, a cloudburst occurred."

There are other correct versions: "For one to work as a mechanic, skill is necessary," "after seeing the doctor, he found his stomach had stopped aching," and "while visiting in Cucamonga, we saw a cloudburst."

Here are five more sentences to which we may profitably apply our second rule, "avoid incomplete constructions." Please complete them on scratch paper and compare with the originals; then turn to the Appendix, p. 176.

Let us repeat that, although you are not being graded on this exercise, practice makes perfect.

Exercise 10. *Avoid Incomplete Constructions*

Clarify these sentences—that is, complete their meanings—by adding the right words at the right places.

1. I remember all my colleagues at the plant better than Fred.
2. The problem is broadened from last year.
3. Entering the laboratory, the damage was seen.
4. The results of these tests showed that
   a.
   b.
   c. (and so on, for perhaps 100 or more words)
5. Encouraged by public approval, the man-on-the-moon tests were stepped up.

Now, as before, let us have a formal test. On a separate sheet, write your corrections of the following 10 incomplete sentences; then turn to the Appendix, p. 177, for the answers.

Exercise 11. *Final for Avoid Incomplete Constructions*

1. This last model sweeper was rejected by every housewife.
2. Having placed a guard at each gate, the danger of security violations is lessened.

3. Working overtime every night, his reputation as a company man was enhanced.

4. Conclusions reached from the previous investigation included:
   a. The weld material showed satisfactory ductility.

5. A good engineer must know how to apply mathematics and be practical.

6. We advised him that to remain silent was insubordinate but to refuse outright was worse.

7. The tone of his talk to us is different yet reminiscent of his former talk.

8. The time passed quickly, reading electronics and Sanskrit.

9. One should not object to summarizing merely on the ground it is repetitious.

10. The workbench was cleared, and the models placed on it, ready for our inspection.

Rule 3. *Emphasize the Main Thoughts*

Not to apply this rule means that we permit our lack of facility with language to rob us of part of what we want to express. Why should we let minor thoughts be emphasized and major thoughts de-emphasized because of our own ineptness? Perhaps because we have never noticed how easily we can repair the damage.

I rewrote the report, thereby making it acceptable to the group leader. This sentence is correct grammatically, but probably incorrect rhetorically. The independent clause "I rewrote the report" could stand alone as a sentence and, since nothing else could, it is grammatically the main thought. But if the writer wants the report-becoming-acceptable to be the main thought, he should make this part of the sentence the independent clause. He should write something like "Because (or "after") I rewrote it, the report became acceptable to the group leader." The "rewrote" part of the sentence has been changed from independent to dependent status; the "acceptable" part, from dependent to independent status. The main thought has been emphasized; the minor thought, de-emphasized.

This space has not been nearly enough. We had to use the other drafting rooms. Here we have two sentences equal in status. Are the thoughts they express also equal in status? Our inept writing has made them appear so. One of the thoughts, if more important than the other, should show this by being expressed in an independent clause. If the second sentence is more important, we would have something like, "Because this space was not nearly enough (or "quite inadequate"), we had to use the other drafting rooms." If the first sentence is more important, we would have something like, "As our use of the other drafting rooms shows, this space has not been nearly enough."

The alkali metals are burned in oxygen or air, and the metallic peroxide is formed. Here again we have two statements expressed as though equal in importance. Connected by "and," they form but one sentence, instead of two. Our problem, however, is the same: if one statement or thought is superior (superordinate), it should be so expressed, and not stated as though equal (coordinate). If we wanted to emphasize the result, we would perhaps say, "When the alkali metals are burned in oxygen or air, the metallic peroxide is formed." If we wanted to emphasize the cause, we might write, "Because metallic peroxide was formed, we know the alkali metals were burned in oxygen or air."

Your order was sent to the wrong department, which is why it was delayed a week. Which did the writer want to stress here, the misrouting or the delay? The first was the cause, the second the effect. If the latter is the more important, it has been poorly expressed, being cast into a dependent instead of an independent clause. It should be, "Because your order was sent to the wrong department, it was delayed a week."

Initiation of the change was accomplished by several methods. Here the author says—but probably does not mean—that several methods were used just to begin making the change. One may have a choice of several methods for effecting some change, but it seems the one used will exclude the others, so that several methods would not be used. If he means that one method after another was tried, he should say, "Successive methods were used before the change was effected."

Does the writer want to emphasize the very beginning of the change? Probably not. If he merely likes big words such as "initiation" and "accomplished," he may mean only that "the change was initiated by several methods." Note what has happened here—one of the polysyllabic nouns was changed into a verb! "Initiation" became "was initiated," a practice that can often be used in simplifying.

Words exist in order to express meanings and should be worked with until they *are* expressed. It is wise to assume that a way exists of saying what one wants to say. This is especially true of technical subject matter, which, after all, requires adequate terminology in order to exist. As the technology develops, so does any new terminology needed. In today's exploding age, as the new sciences ramify and interweave, experts keep coining new terms.

Most of the trouble in expressing oneself, however, lies not in ignorance of technical terms but in lack of ease in (1) clarifying one's meanings to oneself and (2) manipulating sentence elements.

We temper a sentence, as it were, until it suits. Consider the sentence "Words exist in order to express meanings and should be worked with until they *are* expressed." This was first written as "Words exist in order to express meanings and words should be worked with until these meanings *are* expressed." It was seen that the second "words" could be dropped and that, if "they" were substituted for "these meanings," "they" would refer to the noun just preceding it, "meanings." The simpler version would not lose in emphasis; thus these changes were made. Our first rule was being invoked: Use only necessary words. Could it be invoked still further? Why not cut out "in order"? And why repeat the idea of expressing? Why not end up with "Words exist to express meanings and should be worked with as long as necessary"? It was decided that "in order" helped to emphasize the reason for words existing, as did repetition of the "express" idea.

Every sentence one writes should receive individualized care.

The following are five sentences in which to "emphasize the main thoughts." Please correct them on scratch paper, then turn to the Appendix, p. 178.

Exercise 12. *Emphasize the Main Thoughts*

1. A study was carried out, resulting in the selection of three halogens.

2. The design changes are shown in Figures 1 and 2, which stipulated that the hoist mechanism and cask will have an overall height of 15 ft.

3. This section discusses sealing capabilities.

4. Five copies of progress reports shall be submitted every two weeks throughout the investigation, giving the following information.

5. Today we need to think out the right relation between science and society, as never before in history.

Please apply Rule 3 now to the following 10 violations. Rewrite, then turn to the Appendix, p. 180.

Exercise 13. *Final for Emphasize the Main Thoughts*

1. Before the vehicle was modified, the deficiencies were reported.

2. Production of both synthetic gold and diamonds is possible; however, it is not feasible.

3. The manager assembled the supervisors and told them there would be a company reorganization.

4. They had to spend an extra month on the plans, thus losing the $1000 bonus promised them for a speedy submission.

5. Coolant is contained by lead pipes.

6. The nickel cell was enclosed in an evacuated porcelain tube, to prevent collapse.

7. Three scheduled investigations shall be conducted every week, giving the following information.

8. The writer omitted formerly included curves for thermionic systems. He dotted part of the curve for thermoelectric systems.

9. Additional requirements for withdrawing the material will be found in Procedure G-710, and the requirements neither replace nor contradict those of Procedure G-712.

10. A chemical analysis was performed on chips taken from the zirconium control ingot. This analysis appears in Table 4.

Rule 4. *Place All Modifiers Correctly*

Modern students of English help us to understand the importance of this rule. (Modifiers are "words, phrases, or clauses that alter the meaning of other sentence elements by limiting, describing, or emphasizing them.") These scholars point out that, of the several different types of language in the world, ours is "positional," rather than "inflectional." That is, English word meanings usually depend in part on their positions or places in sentences rather than on "inflections," or changes in the letters of the words themselves.

For example, take a simple word like "only," which is not inflected or changed throughout, and place it in the nine possible positions of a nine-word sentence:

```
1. Only I told the foreman what I had seen.
2. I only  "    "      "      "   "   "    "
3.  " told only  "      "      "   "   "    "
4.  "   "   the only     "      "   "   "    "
5.  "   "    "  foreman only    "   "   "    "
6.  "   "    "    "     what only "   "    "
7.  "   "    "    "      "   I only "    "
8.  "   "    "    "      "   " had only  "
9.  "   "    "    "      "   "  "  seen only.
```

We have made nine different sentences by changing the position of a single word. Each of these sentences has a different meaning.

At least three of them can have two meanings apiece. Sentence 2 can mean the same as 5, sentence 7 can mean the same as 6 or 8; sentence 5 can mean the same as 3 or 8.

Users of English thus succeed in communicating only by *ordering* their words with the utmost care. For our first Rule 4 violation, let us consider sentence 2.

I only told the foreman what I had seen. This mistake is understood by all to mean something different from what it says. It means neither what 1 means, nor does it mean that telling rather than thought transference or smoke signals was resorted to as a means of communication. It means 5, "I told the foreman only what I had seen." Though "only" is tagged by the grammarians as an adverb, it here modifies the noun clause "what I had seen" and therefore should be placed just before it.

We have seen how we get different ideas as "only" is moved around in a sentence. We must take care to position correctly several other limiting words: "almost," "scarcely," "even," "quite," "nearly," "just," and "hardly."

I asked him the next time to be more careful. Whereas the preceding violation did not sound odd to us, this one does, thus making it easier to correct. We do so by first identifying the outstanding modifier, that which limits something else in the sentence. It is "the next time." (For diagramming purposes, it may be regarded as a prepositional phrase, "on, or at, the next time.") This is adverbial, the trouble being that it could modify either of the two verbs, "asked" or "to be." It is because we do not know which is meant that the sentence sounds odd. "The next time I asked him to be more careful" does not mean the same thing as "I asked him to be more careful the next time." The first "time" refers to their *discussion* of the carelessness; the second "time" refers to the careless man *performing* the act later. So "the next time" is an example of what is called a "squinting modifier." It looks both ways, back toward "asked" and ahead toward "be." (Squinting modifiers need not, however, be in the middle of a sentence like this; they can squint no matter where they are if not clearly placed.)

Water is taken in through the mouth of the fish and forced over each gill, which is thus their means of respiration. Here the "which" refers confusingly not to "gill" but to the entire independent clause (with its compound predicate) ending with "gill." The relative pronouns "which" or "that" should refer only to individual nouns or pronouns, not to entire complicated statements.

Take this sentence for instance: "Quantitative small-scale studies have been performed to identify the release of iodine-131, which may be invaluable." The word "which" refers *grammatically* only to iodine-131; but *semantically,* or in meaning, to what does it refer? To the release of this element? To its having been identified? To the studies that were performed? Probably the last one.

To correct this sentence, we do what Rule 4 says: place the modifier, the entire clause "which may be invaluable," where it belongs—after "studies." So we have, "Quantitative small-scale studies, which may be invaluable, have been performed to identify the release of iodine-131."

Even a simple sentence such as, "he had to work, which upset my plans," is wrong because the "which" refers to the entire independent clause "he had to work." We would correct it either by saying "his having to work (these four words function as a single noun) upset my plans" or "he had to work, a *fact* that upset my plans." Here we import a noun, "fact," to serve as the simple referent for "which."

The principal matter before the board, which brooked no delay, was nevertheless tabled for a week. Here, despite the misplacement of "which brooked no delay," we know what is meant. Since we mildly despair of correcting the order, we tend to do nothing. But such an error-accumulating attitude can soon make of a report a thing of sorrow and a pain forever.

Let us see what can be done. Grammatically the "which" refers to "board," but in meaning it refers to "matter"; the entire expression, "which brooked no delay," is an adjectival clause modifying "matter." Appearing where it does, it violates our rule to "place all modifiers correctly."

Instead, we could say:

1. "The principal matter before the board brooked no delay, yet was tabled for a week."
2. "Although the principal matter before the board brooked no delay, it was tabled for a week." (The "it" more clearly refers to "matter" now than did the original "which.")
3. "The principal matter before the board, although brooking no delay, was nevertheless tabled for a week."

Any of these is clearer than the original.

It is valuable to regard every sentence as belonging to a surprisingly large family of closely similar sentences. The out-and-out black sheep of the family violate the various sentence rules; of the members in good standing, one sentence is superior to all the rest and should be chosen!

By the way, which of the preceding sentences do you prefer—1, 2, or 3? Our vote goes to 2.

I found the trouble over the weekend that rendered my microscope useless. Here "over the weekend" squints. Placed as it is it means that the uselessness of the microscope occurred over the weekend. If we want to say that the trouble was discovered then, we place the phrase where it will clearly indicate this: "Over the weekend, I found the trouble that rendered my microscope useless."

Let us now correct five more sentences that infract our rule, "place all modifiers correctly." As before, write your amended sentences on scratch paper, then turn to the Appendix, p. 182, to compare your final versions with the ones there.

Exercise 14. *Place All Modifiers Correctly*

1. I only told my supervisor the truth.

2. He bought the chemical over the weekend that was needed to complete his experiment.

3. The man lives in a four-room house with his son, which he rents for $150 a month.

4. Because ether volatilizes when not in use it should be stoppered.

5. The "expert" was asked kindly to resign.

The following is a formal test to correct 10 violations of Rule 4. Please write your corrections on separate paper, then see Appendix, p. 183.

Exercise 15. *Final for Place All Modifiers Correctly*

1. Newspaper headline: Stereo Set, Cash
　　　　　　　　　　　　　　Taken by Burglars
　　　　　　　　　　　　　　From Beauty School

2. I found the report that praised him in the wrong folder.

3. In machining there are two distinct operations done by experienced workmen called reaming and broaching.

4. You told me once she was a champion typist.

5. The foreman promised as soon as possible to review the complaint.

6. I had a split second to just swerve the car to the right.

7. He promised to order the printing as we were leaving.

8. He weighed the concentrate on the scales that had been carefully prepared.

9. An empty process tube, a dummy fuel element, and two in-core instrument thimbles were successfully washed, which had been in the core during hot circulation tests.

10. At least one suggestion prize announcement is made in every edition of our weekly newspaper almost.

Rule 5. *Use Similar Constructions to Express Similar Situations*

Language is of course used to express thought rather than conceal it. One of the outstanding features of the world is the fact of similarity; that is, that certain objects resemble one another, and can therefore be placed in classes. Our minds grasp these relationships, observe many instances of them, and are thus able to classify scientifically, to recognize, to feel at home in the world. So we have similar thoughts because they are aroused by similar entities.

For example, Einstein had certain *qualities* such as initiative, persistence, daring, and humility. These are similar in being traits pertaining to a person. We *express* their similarity grammatically by terming them all nouns in a sentence in which they are attributed to Einstein. We carry similarity of the facts over into a similarity of grammatical structure. That is, to help the reader recognize a similarity in nature, we devise a similarity in the sentence structure that describes it. The sentence "parallels" the reality.

The happiest situation grammatically, in regard to the expression of similarity, is the one in which such parallelism is possible. In the following example, note how parallelism can extend to the various parts of speech.

Noun. Einstein had initiative, persistence, daring, and humility.
Adjective. Einstein was acute, curious, determined, and dedicated.
Verb. Einstein questioned, analyzed, persisted, and finally triumphed.
Participle. Einstein faced "negative" facts fearlessly, insisting on the truth as he saw it, applying each theory thoroughly, and abiding by the results.
Gerund. Einstein won his success by asking basic questions and by considering odd answers.
Infinitive. Einstein liked to propose, to reason, to test, and to verify.
Phrase. Einstein amazed the world first with the special theory of relativity, last with the general theory of relativity.
Clause. Einstein believed that Newton was wrong in some of his basic ideas but that he was right to believe in a God.

We can easily spoil these thought-assisting parallelisms by writing unskillfully. We could say, "Einstein had initiative and was persistent too, along with daring and his being humble," or "Einstein liked to propose and was fond of reasoning, also liking to test and verify." In fact, since it is always an effort for even the greatest writers to move from thought to expression, we may expect in all of us an initial *un*parallelism. Our similar thoughts are not born neatly parallel, grammatically. We must be willing to work at achieving parallelism.

Similarity seems to exist in varying degrees. A grown horse is similar to a grown elephant in that both are quadrupeds, but the horse is *more* similar to its own colt, even in size. At one end of the scale similarity can be neatly expressed in various easy parallelisms of expression. At the other end similarity is expressed minimally in what grammarians call "agreement." This involves pronouns and verbs, as follows.

Pronouns must agree with (i.e., be similar to) their *ante*cedents in these three respects:

Number. *Example:* "Every machinist went out on strike because his (not their) demands were ignored." The antecedent, "Every machinist," is singular; therefore the pronoun "his," which links that person's demands, should also be singular.

Person. *Example:* "George continues to study electronics because it keeps him (not you) in touch with the world of today." We maintain consistency in person between the pronoun "him" and its antecedent "George," both of which are in the third person. "You" would be second person. Otherwise the meaning is obscured.

Gender. *Example:* "Mr. Holbrook is the only executive now in the plant who (not which) is directly responsible for the program." Since "Mr. Holbrook," the antecedent here, is masculine, the pronoun referring to it should not be of neuter gender ("which"). "Who" and "whom" are either masculine or feminine in gender.

Verbs must agree with (i.e., be similar to) their subjects in the following two respects. (Verbs have no gender.)

Number. *Example:* "There are (not "is") a chair and typewriter in the room; now sit down and write your report." The "chair and typewriter" constitute a plural subject: Therefore the verb should "agree" by being plural; that is, the irregular verb "to be" changes from "is" in the present tense singular to "are" in the present tense plural.

Person. *Example:* "Neither you nor I accept (not accepts) this decision as final." In the "neither . . . nor" type of sentence, the verb agrees with that one of the two subjects that just precedes it, "I" in this case.

These five situations involving agreement between pronouns and their antecedents and between verbs and their subjects are, then, contained in Rule 5. The five rules *prescribe* that certain similarities among facts in the "outside" world be expressed by certain similarities in word usage. The language alternatives preexist so that there is, for example, a word "his" to use when we need it, instead of just "their."

Let us discuss five more examples of Rule 5.

The chief chemist warned his superiors of insufficient materials in the factory, and that they were also in short supply nationally. This is an example of a sentence in which we sense an unexpressed parallelism. If we persevere in analyzing until we have spotted it, and then express it, we begin to form a valuable habit. We note that shortages exist not only locally but nationally. These two situations should be expressed similarly in the sentence but they are not: one expression, "in the factory," is neat enough but the other, "and that they were also in short supply nationally," is diffuse.

What would be the grammatical parallel for "in the factory"? Why, "in the nation." So we join the two prepositional phrases by "and." Note that if we merely said "nationally," though this would still be better than the original "that . . ." clause, the parallelism would be incomplete. The more we say it to ourselves, the more we like it: "The chief chemist warned his superiors of insufficient materials in the factory and in the nation."

Why was the sentence strung out originally? Probably because the writer liked the phrase "in short supply" and wanted to use it. But in doing so he was repeating the idea, since "insufficient materials" had already been mentioned. By insisting on sounding like an economist, he muffed his chance to effect a neat parallelism.

While the motor is idling, and with its spark retarded, it operates quietly. Here is a more easily sensed parallelism, that of the idling motor and the retarded spark. Which type of expression shall we change into the other? That is, shall we make the motor clause a prepositional phrase or the spark prepositional phrase a clause?

Let us try them both and see which we prefer. The first change would give us "With the motor idling and its spark retarded, it operates quietly." The second change would give us, "While the motor is idling, and while its spark is retarded, it operates quietly." We prefer the former.

The city has still not increased their taxes. The referent of the pronoun "their" is "city," which, being singular, dictates the same form to the pronoun. "Their" must therefore become "its."

He lists his objections to the proposal, but its good points are not even referred to. The lack of parallelism here consists in having one of the two independent clauses "He lists his objections to the proposal" in the active voice, but the second independent clause ("but its good points are not even referred to") in the passive voice, although the two assertions are similar in character. Although "he" is involved in both of the clauses representing the two similar situations, "he" is the subject of only one of the clauses. If "he" is made the subject of both clauses, thus putting them both in the active voice, the reader will be helped to grasp the similarity of both situations. So we might have, "He lists his objections to the proposal, but does not even refer to its good points."

He is in poor health, and needs teeth, house, and car repairs. This sentence should reflect the fact that health and teeth belong in one group pertaining to the man's person, with house and car repairs in another group pertaining to his property. So we might say, "Besides being in poor health, his teeth require care; his house and his car also need repairs.

Now write your own answers to Exercise 16, then turn to the Appendix, p. 184.

Exercise 16. *Use Similar Constructions to Express Similar Situations*

1. We gave each of them a chance to protest, and then they could file a petition for transfer.

2. He is one of those sensationalists who has given research a bad name.

3. First consider the origin of this theory and then how it has developed.

4. Mixtures of potash and sulphur require a heavy blow to explode them, whereas you can explode compounds of fulminate of mercury by giving them a hard look.

5. He proceeded methodically, and the necessary steps were taken in proper order.

Let us now correct 10 more violations of Rule 5. As before, please write corrections on separate paper, then compare with answers in Appendix, p. 186.

Exercise 17. *Final for Use Similar Constructions to Express Similar Situations*

1. I deal with the discount houses because of their lower prices and their stock is larger.

2. The Forbes Company plans to extend their holdings.

3. Darwin was admired by fellow scientists, but the clergy hated and slandered him.

4. The laser is one of the most interesting inventions that has ever been made.

5. Let us first review the cause of the trouble and how it was handled.

6. Our plant has 20,000 workers; we are not Fascists, but one must provide moral education here, or you will turn out Communists.

7. The stipulated task of cutting, drilling, and polishing were performed rapidly.

8. If anyone objects, I shall discharge them.

9. Businessmen face knotty problems in estimating costs and particularly gauging demand.

10. Gallium can be melted like this, but you cannot melt steel that way.

Rule 6. *Employ Precise Terms*

This rule would require a great essayist to do it justice. He would show not only its scope and importance, but also the difficulty of applying it.

Unfailing preciseness must remain an ideal. But the technical man can develop habits—maybe hobbies—that will steadily increase his mastery over the great world of words. We should first realize that logic is all with the reader who insists on clarity of expression and with the writer who keeps struggling to achieve it. If the latter shirks, thinking, "Oh, that's good enough—they will know what I mean anyway," he is piling up reader-frustrations that may spell his failure as a writer and even as a professional man. Why should readers have to supply a string of inferences along the way, like, "Well, he said this, but I guess he means that"? Technical writing and editing should produce unmistakable meanings.

Consider this directive:

"The policy of the ——— (military department) with respect to the use of oleomargarine is that it should be used in a ratio to butter in a proportionate amount to be predicated on its acceptability to the men." (36 words)

When detective work is done on this mysterious policy, it becomes:

"The ——— wants *as much* oleomargarine *substituted* for butter as service men *will accept*." (14 words)

Our italics are to show how simple the clarifying words can be—usually it is not the big words, but the little words, that supply the needed precision.

Sometimes the more hopeless a piece of writing is, the quicker we can see what is wrong. Take the foregoing as an example. When we try to figure out what it means, we see that the military department *wants* something. So we do not need "The policy of the ——— with respect to . . . ." The *use of strong, simple* verbs is a partial cure for sick writing.

What is wanted, moreover, is not just oleomargarine, but some practice associated with it. As we wrestle with the meaning of "it should be used in a ratio to butter in a proportionate amount . . . ," we suddenly realize that the writer is merely thinking about *substituting.* Unfortunately, this everyday word was not in his working or acitve vocabulary, so he was forced to *write all around it.* This demonstrates how important a good vocabulary is for clarity. Vocabulary is the capacious mother of precision.

As for "to be predicated on its acceptability to . . . ," remember what we just surmised, that simple, healthy verbs can often cure bedridden prose. How can one get a strong verb out of this last quote? It is often done by verbalizing a long, stuffy noun, and we have one here—"acceptability." Verbalizing it gives us "accept."

The main virtue of the 14-word rewrite is not that the wordage was reduced by 62 percent, but that clarity was achieved. The directive can now be understood. Originally, one could not *see through* the words to what was meant.

Precise terms are not necessarily "concrete" terms, words referring to physical objects. Precise terms can just as often be "abstract," words dealing with qualities and relationships. In the directive just analyzed, one of the key words, to "substitute," is abstract as it refers to a kind of act. Being precise in one's writing has to do only with using the word that says exactly what one means, the "specific" word. Nor is this double-talk; it is single-talk.

To help us express our desired meanings by using just the right words, we have prepared 1400 "aura words," chosen as carefully as we could devise for *supplementing* the vocabulary of a technical man. They begin on p. 188 of the Appendix. Everyone should have these words in his active vocabulary. To assure this, the best beginning exercise is to check all one's opinions as to the 1400 meanings with the dictionary.

Let us now discuss five violations of the rule to employ precise terms.

The design of this equipment has carried with it the view of ease of maintenance. How can a design, an inanimate object, *have* a view, or *carry* one? Since the verb "to carry" does not help out here, let's scrap it. We need a working verb. Can we perform our trick of finding it by converting one of the nouns in the sentence? (The English language is filled with such instances of "functional shift," the changing of one part of speech into another.) Yes, again there is a noun made to order—"design." But we must make "equipment" the subject. "This equipment was designed for ease of maintenance." The functional shift was from the use of "design" as a noun to its use as a verb.

It hardly felt as if it were heavy enough. "Hardly" is too colloquial. Furthermore, it is placed as though it modified "felt," whereas it modifies "heavy enough." The word "scarcely" is superior and, when better placed, would give us, "It felt scarcely heavy enough." But this is still too imprecise for a technical man. He would prefer to say, if fact supported him, "Its weight was inadequate"—these four words are better than the original nine.

Mrs. Green said her husband never had a mother. Didn't Mrs. Green know the word "know"? In her statement to the social worker (most of the faulty sentences in this book are "real"), Mrs. Green should have said that her husband never knew a (or his) mother.

Mr. Parker had eaten his last meal ticket. Here is another funny sentence from a social service report. He should have said "used up" instead of "eaten." Perhaps the reader can further improve the sentence.

The effort concerned with this test was carried out. An engineer wrote this. The technical editor must decide if it should stay in the report. If he thinks not, he must deftly correct it, then just as deftly make sure that the engineer-author approves. One feels that he wants to say something very simple here but has difficulty in manipulating the words. The result is overwriting and complicated jumbles. What does the writer mean by "effort" and "carried out"? Was this test particularly difficult? Was the "effort" itself carried out on a stretcher, exhausted from trying to finish the project? When the author decides just what he wants to say, he should write it as simply as possible. He may mean, "An effort, unsuccessful as yet, was made to perform the test," or merely, "The test was performed."

Let us now correct five more sentences that violate the rule to "employ precise terms." As before, write your amended sentences on scratch paper, then compare your versions with those in the Appendix, p. 196.

Exercise 18. *Employ Precise Terms*

1. This work is to support the main experiment.

2. The mixture was more or less saturated with oil.

3. If the activities at your locality feel strongly that the requirements should be revised, they may recommend through proper channels their implications for the revision.

4. The control of the fuel cask operation would be remote, and would be viewed through shielding windows.

5. In writing any report, the readers have a different familiarity with the report subject.

Let us now correct 10 more violations of this rule. Write each improvement on separate paper, then compare with Appendix, p. 197.

Exercise 19. *Final for Employ Precise Terms*

1. This computation has been performed under another project.

2. The enlightenment of this experience has provided a firm basis on which to write equipment specifications.

3. While the metal was rolled out to the desired length, it became much too brittle.

4. The can was designed for the use of rods.

5. Checkout is not yet complete, but first results are in a direction to provide greater agreement with experiment.

6. A man's work should be referred to his peers as a standard.

7. The preparation of the above referred proposal pointed out several significant factors regarding the concept and resulted in several conclusions being drawn.

8. The present design effort was concentrated on providing a more gentle transition.

9. This was done in order to upgrade the conditions for testing carbides.

10. A new age dawned when they finally invaded space.

We have now discussed in detail all but one of the rules for writing effective sentences. The complete set is:

1. Use Only Necessary Words
2. Avoid Incomplete Constructions
3. Emphasize the Main Thoughts
4. Place All Modifiers Correctly
5. Use Similar Constructions to Express Similar Situations
6. Employ Precise Terms
7. Try Habitually to Improve Sentences

It is probable that you have not yet learned these rules. But remember that "practice makes perfect" applies not only to the athlete but to all learners. When one has failed to learn, it is usually because he has failed to review often enough.

Rule 7. *Try Habitually to Improve Sentences*

Next to repetition, we think the best of the learning practices is to present a subject in a different way. Let us do this. So far our problems have been *directed*. We have known what we were going to do: for instance, emphasize the main thought in a sentence.

Our last rule, "try habitually to improve sentences," simply tells us to apply all the other rules. This is an *undirected* rather than a directed rule; it is the psychological rule, applicable to writing in general.

Since a faulty sentence is apt to show more than one fault, the analyst learns not to give up too soon; he becomes dexterous in using the rules; he keeps evaluating the sentence until it passes inspection. If you do not quite know these rules yet, drilling on them in this wider, undirected way should help.

Applying the last rule will train you to recognize the various faults in sentences without too much trouble. You must learn to persist in your correcting until you have ironed them all out—not just the glaring one.

The five following examples of combined faults will call upon all your knowledge.

The scope of the report encompasses the steam-producing facility only. Consider one rule after the other, to see which one applies. With practice, this will become automatic.

Are there unnecessary words? Sometimes these can be spotted instantly; often, however, it takes time to detect meanings that repeat or overlap. Such a case confronts us: "encompasses" and "scope." Since the first word is a verb and the second a noun, this may be difficult to clear up. Let us drop one of them, "scope." Then we have, "The report encompasses the steam-producing facility only." This sounds better, a fact that is rather surprising since we have changed the subject. But we have changed the subject before now, to advantage; we should not hesitate to do it when sentences seem stuffy.

So far, then, we have found one rule violation, a violation of "use only necessary words." But are we quite content with what is left? Let us regard it with a gimlet eye: "The report encompasses the steam-producing facility only." What is this four-syllable "encompasses"? It sounds pompous; is it indeed the specific word needed here? We can either think hard about what "encompasses" means and refer to a dictionary or we can order our computer-brain to print out to consciousness any synonyms that it can from our memory bank of words, our vocabulary. Suppose we do both.

The first meaning given for *encompass* is "to form a circle about, or enclose." This is not what we mean. The second meaning is "to envelop or include." Did the author mean to say, "The report includes the steam-producing facility only"? No; or he would not have said "only." The two words go in opposite directions: "includes" lets something in, "only" shuts something out. The third meaning is "to bring about, or accomplish." This would not sound correct either. We have to conclude that "encompasses" just misses all around.

So let us try the memory bank, saying the sentence over slowly, hoping that the right word will print out. "The report concerns the steam-producing facility only." We repeat it. Yes; what the author meant to write was "concerns"—a word that makes "only" sound sensible. So another rule—"employ precise terms"—was violated in this short sentence.

But what about "only"? We remember that modifiers should appear close to the words they modify or belong to. This is an adverb modifying "concerns," yet it appears five words away. Can we not do better? Yes: "The report concerns only the steam-producing facility." A third rule had been violated.

Is it worth going through all this trouble to improve just one short sentence? Well, it is not as difficult as it still may sound; to repeat, practice and habitude are the answers. The improving thoughts are part of one's rapid stream of consciousness; they move right along. We started with "The scope of the report encompasses the steam-producing facility only" and ended with "The report concerns only the steam-producing facility." The engineer said, "Yes, that's what I meant." We hope he felt a bit sheepish. *We* had to impose Rules 1, 6, and 4 to get the sentence right.

The instructor is experienced in jungle fighting, which adds interest to the course. Which rules does this sentence infract? How about 4? Remember the example beginning "Water . . ." (p. 60)? Here is a similar case. Grammatically the "which" refers only to jungle fighting, but it was probably intended to refer to the instructor's experience in it. If so, the sentence should be recast so that "which" can refer to "experience" as the first noun preceding it. We would then have, "The instructor has jungle fighting experience, which adds interest to the course."

But something else is wrong. The fact of a man being experienced in such fighting is of a certain interest. But that students *benefit* from this is more important. Yet "which adds interest to the course" is a dependent clause, whereas the less important fact is the major clause. So the rule to "emphasize the main thoughts" is also violated.

But note that if we correct this second fault, the first one disappears! "The instructor's experience in jungle-fighting adds interest to the course." No "which" exists to cause trouble. We did not plan it this way; it just happened.

This has been carried out. No doubt what "this" referred to, if the sentence were left in context, would be clear. But not clear enough! A noun expressing the referent should be added: "this project," "this plan," whatever it was. So the rule to "avoid incomplete constructions" was violated.

In addition, "carried out" is not precise enough (Rule 6). Carried out with the trash? No. "This plan has been effected," "finished," "concluded," "put into effect"—these are all better than "carried out."

The tests during the past month have shown improved reliability in the test data measurements and more consistency. To begin with, "improved reliability" and "more consistency" seem to constitute two results meant to be the direct objects of the verb "have shown." If so, why are they separated? Could it be because "in the test data measurements" modifies only "reliability?" Hardly; the measurements have no doubt shown both of these results. Why should we not say, "The tests during the past month have shown improved reliability and more consistency in the test data measurements." Rule 5, "use similar constructions to express similar situations," has been applied.

But we do not like the "test" idea to appear twice. Also, if we have a past-tense verb and say "during the month," isn't the *word* "past" just in the way? If "consistency" is placed near "reliability," can't "improved" modify both nouns, thus obviating "more"? (Rule 1.)

Finally, why have "in the test data measurements" at the tail end? (That is the way it happens; we begin to heartily dislike the little ineptitudes.) Such a change would increase the emphasis (Rule 3).

How about "Test data during the month have shown improved reliability and consistency"? The writer said yes, that is what he meant.

A paper on the subject of eutectic alloys for metals has been written, and approved for publication in Acta Metallurgica. What rules are violated? To have these rules well-learned and ready to apply is probably the quickest way to begin rewriting. If we can drop "the subject of," we invoke Rule 1. After we have made the rules a part of us, can we not use them without bothering to name them? Of course.

What about wasting words in saying the article has been written? Can't we assume it was, if approved? So we can drop "written, and."

Notice that if we use Rule 1, we will also have applied Rule 3. Here is a second instance of accomplishing something we had not expected: we shall have emphasized the main thought. We now have, "A paper on eutectic alloys for metals has been approved for publication in Acta Metallurgica."

The following are five more examples of variously combined faults. Please identify, as well as correct, all the types of faults committed. Then consult Appendix, p. 199.

Exercise 20. *Try Habitually to Improve Sentences*

1. The objective of this project is the design of a fuel element also to prepare a safeguards report.

2. Perhaps this poor writing is due to the press of schedules, but there surely should be some improvement along these lines.

3. There appears to be two problem areas in regard to adjusting the variable orifices.

4. Hi-Tec is limited to a maximum temperature of 1000°C due to pyrolytic decomposition at high temperatures.

5. His most biting comment is that young men who are helping fight a war should be subjected to such a humiliating experience.

On separate sheets, please write your corrections of the 15 following sentences, which instance all of the faults discussed in the text. Then check with Appendix, p. 200.

Exercise 21. *Final for Try Habitually to Improve Sentences*

1. The discrepancy between theory and experiment noted in the last previous quarterly report has been carefully examined.

2. The problem of high gas solubility and its effect on heat transfer is unknown.

3. The detection of Fe, Cr, and Ni in thin film deposits and of slight surface irregularities in the hot-leg imply corrosion may have occurred.

4. The reason he failed was he could not take criticism without losing his temper.

5. Chemical analyses for the alloyed ingots appear in Table 4. The alloyed ingots were analyzed for carbon and nitrogen and the alloying elements.

6. The study infers that material erosion and corrosion data can be extrapolated to larger systems.

7. The new employees will be here tomorrow and processed the day after.

8. Components have been operated only for testing purposes, resulting in low maintenance requirements.

9. Having at last found the solution, the problem was solved.

10. The purpose of these experiments on improving the perfection of UC crystals is to prepare crystals that are sufficiently flawless to permit the use of X-ray microscopy.

11. As a direct result of the engineering mock-up work, the following are problems to which feasible solutions are apparent.

12. The hot rolling schedule is presented in Table 7. As the rolling schedule indicates, the rolling temperatures varied with the alloy content.

13. By this procedure it was established that defects could be removed.

14. If the chamber is sealed, it will greatly increase the pressure.

15. I always have and will always keep trying to meet my deadlines.

PROOF OF THE PUDDING—RULES APPLIED TO "HORRIBLE EXAMPLE"

We are now prepared to apply our knowledge of sentences to the unit of writing found everywhere in industry, the "IOM" (interoffice memo), "AVO" (avoid verbal orders), or whatever the major type of short communication among employees and departments is termed. These memos are frequently so badly written that "beware of the written word" has become a cynical motto of plant life. In one aerospace firm with 15,000 employees, for example, one in five executives admits to ignoring memos because they are poorly written or downright uninterpretable.[1]

Following is a "real" directive that is almost meaningless. Suppose your superior asked you to rewrite it?[2] Let us note as many faults as we can before detailing our analysis.

Here is the poorly written directive.

> It is generally acknowledged that proper time charges have been thoroughly confused within the division. In an effort to reduce the amount of legwork incurred by designers within the unit, I have asked Mr. Walden to establish the responsibility of giving time charges for new projects to L. Harris. This has been agreed to; therefore, the Unit Secretary shall maintain an up-to-date time sheet concurred to by Harris of all charges including overhead. Any unresolved questions should be referred to immediately to Harris.

This is almost, but not quite, hopeless. We must start by *identifying errors*. Then we *cultivate articulateness* by making ourselves express these errors in words. We can not correct what we haven't expressed to ourselves as well as we can correct what we *have* expressed to ourselves. First, then, what errors do we see?

[1] Graham B. Bell and B. J. Hoffman, "Attitudes Toward the Interoffice Memo as a Communication Device," *Journal of Social Psychology*, 15–21 (June 1965).

[2] "Skill in rewriting for clarity at reader's level" was considered the most important of eight qualifications for technical writers, in a survey of 160 industrial firms. John A. Walter, "Education for Technical Writers," *Society of Technical Writers and Publishers Review* (January 1966).

1. In a directive, is it necessary to say something is "generally acknowledged"?

2. What does or could "confused" mean here? No less than three distinct things: that time charges have been (a) made incorrectly, misstated, (b) made correctly but misunderstood, or (c) jumbled up physically, disordered. Until we know which of these is meant, the directive will not direct.

3. Why should "in an effort to reduce" be used? This suggests difficulty of solution rather than that the problem of making time charges has been solved. The latter point should be stressed rather than the former because we *are,* presumably, announcing the solution and enlisting everyone's cooperation in a new procedure.

4. What is this "legwork" business? What is the connection between making proper time charges and reducing designers' legwork?

5. What does this mean: "I have asked Mr. Walden to establish the responsibility of giving time charges for new projects to L. Harris"? Does it mean Mr. Walden is to tell people they simply must give their time charges now to Mr. Harris? Probably not. It seems to mean that Mr. Harris has just been *designated* or *empowered* to allocate time charges for new projects. The writer did not have "designated" or "empowered" in his working vocabulary so he wrote around it by saying such a roundabout thing as "establish the responsibility of . . . for . . . to . . ." This shows a lack of the so-called specific word and how a limited vocabulary hampers one in writing.

6. The use of the pronoun "I" is bothersome. The directive involves three persons—I, Mr. Walden, and Mr. Harris. But it bore the author's name; therefore the identity of the "I" was known. So "I" could have been dropped and the "person" of the writing changed from first to third, thus leaving only two individuals to keep in mind. Also, third-person communications in an office are more businesslike and modest.

7. The "this" of "this has been agreed to" is doubly confusing. As we said earlier "this" is used properly to refer to a noun or pronoun, not to an entire situation. Whenever we use "this," "that," or "which" to indicate a situation, we should add properly identifying and reminding words such as "situation," "state of affairs," and "plan of action."

In the second place, the situation to which "this" refers was never clearly stated. (Remember the three possible meanings of "confused.") The use of "this" makes the doubly irritating assumption that we know (a) what has been agreed to and (b) what caused the original confusion.

8. "Has been agreed to" is confusing. An order or directive does not require agreement, or if it does the agreement has preceded the order and need not be announced afterward. Reference to agreement is wasteful;

it does nóthing except make the reader wonder if he understands what he is reading.

9. "Concurred to" should be "concurred in."

10. The name "Harris" should be preceded by its modest title, "Mr.," since the "Walden" name is. Or both "Mr.'s" should be deleted. The editor's basic rule is consistency.

11. The chain-of-command understanding already established by this directive is disrupted, and the reader is made to rethink the situation. He thought that Mr. Harris was being given a certain time charge responsibility. Now he reads that a "Unit Secretary" is to maintain a time sheet—to which poor Mr. Harris will have to agree! Since our experience tells us that secretaries do not dictate to principals, as Mr. Harris apparently is, we conclude that the writer again means something different from what he is saying.

12. In the last sentence, ". . . referred to . . . to Harris" is both ungrammatical and confusing. The preposition "to" should be followed immediately by its proper object, "Mr. Harris," and the second of the two "to's" must be scrapped.

13. There is something wrong with the phrase "unresolved questions." Questions are questions *because* they are unresolved, at least by somebody. Certainly in this case, in which we are talking about any questions that may be asked by the employees in an entire division, we can be sure that the questions will be sincere—that is, as yet "unresolved." So this word is unneeded.

14. We note that the way in which this final sentence is written vaguely annoys us. Since we know that professional writing rests on an analytical foundation, we stubbornly ask ourselves why we are irritated. We conclude it is the "should be referred to" part that does it. This is written in the passive voice, stating that something is done *by* something or somebody else. What other way is there to say it? Why, that somebody or something *does* something—active voice. We have already felt that Harris's role was mistakenly minimized in favor of the "Unit Secretary." Instead of going along with this and saying that questions should be given to Mr. Harris, why not say he should answer them? We thus emphasize the important point.

Let's draw a deep breath after noting those 14 mistakes. Are you wondering whether correct writing is worth all this trouble? If so, be reassured: we have just seen that mistakes in plant and office communication can be verbalized. True, it is difficult to detect these mistakes and analyze them in order to obtain clarity. This is, however, precisely the kind of thinking that the writer bent on competence willingly trains himself to do.

For here is the exciting corollary: *if a mistake can be verbalized, it can be corrected; if the wrong words are identifiable, the right words are substitutable.* The circulating coin of poor writing has its obverse—good writing. The writer who permits himself to be trained soon *stops* choosing the wrong words.

Here is what we started with:

> It is generally acknowledged that proper time charges have been thoroughly confused within the division. In an effort to reduce the amount of legwork incurred by designers within the unit, I have asked Mr. Walden to establish the responsibility of giving time charges for new projects to L. Harris. This has been agreed to; therefore, the Unit Secretary shall maintain an up-to-date time sheet concurred to by Harris of all charges including overhead. Any unresolved questions should be referred to immediately to Harris.

Here is how the author might have written it, thus saving many man-hours of argument over what it meant.

> To correct the improper making of time charges within the division, Mr. Walden has empowered Mr. Harris to
> (a) see that time charges for new projects are made properly,
> (b) have the Unit Secretary maintain an up-to-date time sheet of all charges, including overhead, and
> (c) answer all related questions from personnel.

The rewrite is simple and direct. Notice that for all of the following,

> It is generally acknowledged that proper time charges have been thoroughly confused within the division. In an effort to reduce the amount of legwork incurred by designers within the unit. . . .

has been substituted,

> To correct the improper making of time charges within the division. . . .

For this,

> I have asked Mr. Walden to establish the responsibility of giving time charges for new projects to L. Harris

has been substituted,

> Mr. Walden has empowered Mr. Harris to. . . .

The rest of the original directive, starting with "this has been agreed to," was rethought and rewritten into the *display listing* (setting out by itself on the page) of points a, b, and c. Each of these begins with a verb, "see," "have," "answer," thus presenting similar situations in similar ways for the sake of clarity.

Notice that the contraction into an understandable communication has been made possible by use of *specific words* such as "correct," "improper," "empowered," "answer," "related," and "personnel." For lack of these everyday words, one simple little situation after another went unexpressed.

Even in our rewrite, "time" and "charges" appear three times each. Is this necessary? We left in "improper making" and "made properly" because it creates a kind of antithesis or balance to emphasize a main thought of the directive, namely, its purpose.

Let us emphasize that the reason for working slowly and painfully through all the pertinent points discussed is to build up a set of skills for avoiding mistakes. One soon begins to think that extra thought before writing. It often becomes as easy to write well as it was to write poorly.

Since this chapter has 14 special exercises, no "final" is suggested.

III

How to Write Strong Paragraphs

A person can greatly improve his writing by taking the construction of paragraphs seriously. How? He writes them by always thinking deliberately about and providing for topic sentences, principles of development, and transitional phrases. Through extensive work in paragraph rewriting, we see that all our sentence knowledge must also be brought into play. We discuss many examples of the paragraph in preparing for the two self-administered tests that are answered in the Appendix.

GENERAL REMARKS

A paragraph is, ideally, a group of sentences expressing a single topic or subtopic; it is built around one dominant idea. Although it rarely does, it may consist of but one sentence. Paragraphs are units midway between outlines and the sentences that compose the paragraphs.

How long should a paragraph be? That depends upon three factors: the style of writing, the subject of the paragraph, and the scale upon which the entire subject is written. As for style, a popular article style might divide this paragraph into three paragraphs, to make it easier to read; whereas a scientific treatise would not. As for subject, suppose it were, "The process involves three quite simple operations." We would expect the paragraph to be long enough to cover them. Yet if the scale of the writing were, say, an operations manual, the three simple operations might each require many paragraphs, pages, or chapters of explanation.

The principles of unification used in writing paragraphs are frequently not clean-cut and are usually employed in combination. Yet the paragraphs hang together quite well, which is surprising when one reflects that paragraph-forming is scarcely a deliberate activity of the mind.

We must distinguish between a paragraph that is such in form only, that is, it merely begins with an indentation, and one that reveals, as it should, a unifying topic.

TOPIC SENTENCES

Most paragraphs have so-called topic sentences around which the other sentences group themselves. The topic sentence, wherever placed, unifies the paragraph.

1. *Physics is tied to all other sciences.* In astronomy and geology results are interpreted according to physical principles. Chemistry makes increasing use of the results of modern physics. Whether or not life is explainable in terms of physical principles alone, living things are made of matter, and the biologist must therefore treat physical knowledge seriously.

The assertion made by the topic sentence is illustrated by showing the relationship of four other sciences to physics.

Now examine this paragraph:

2. *The facilities of the hydraulic laboratory* include a channel 20 ft long, 1 ft wide, 2 ft deep, to be supplied at the rate of 2 gal per sec from a tank 3 ft square and 10 ft high. The tank was designed to be used for plunging experiments. There is also a high-speed water table 3 ft wide and 10 ft long, fed at the rate of 4 gal per sec. Next, a small power plant was proposed, which will consist of a 2 hp pelton wheel, electrodynamometer, and associated regulatory equipment. The facilities include provision for circulating three fluids of different densities and viscosities through a selection of pipes under controlled temperature conditions.

Oddly enough, this paragraph has only a *topic phrase.* Now consider the following example.

3. *The difference between common and scientific knowledge is roughly analogous to differences in standards of excellence set up for handling firearms.* Most men would qualify as expert shots if the standard of excellence were the ability to hit the side of a barn from a distance of a hundred feet. But only a much smaller number of individuals could meet the more rigorous requirement of consistently centering their shots upon a three-inch target at twice that distance. Similarly, a prediction that the sun will be eclipsed during the autumn months is more likely to be fulfilled than a prediction that the eclipse will occur at a specific moment on a given day in the fall of the year. The first will be confirmed should the eclipse take place during any one of something like 100 days; the second will

be refuted if the eclipse does not occur within something like a small fraction of a minute from the time given. *The latter prediction could be false without the former being so, but not conversely; and the latter prediction must therefore satisfy more rigorous standards of experiential control than are assumed for the former.*[1]

The topic sentence of this paragraph is evidently the first one. The final sentence learnedly confirms what the first sentence says, that scientific knowledge is more exact than common knowledge. In long, well-written paragraphs one may find the topic sentence recurring at the *end* in this way, rewritten. It clarifies.

A succession of paragraphs each starting with a strong topic sentence could be monotonous. Therefore, such sentences may occur at other places in the paragraph.

Also a paragraph which is transitional in nature may quite properly lack a topic sentence. The function of the transitional paragraph is simply to connect the paragraphs preceding and following it.

PRINCIPLES OF DEVELOPMENT

After noting that paragraphs are unified by their constituent topics, we next observe that paragraphs are developed by the use of the following: details; cause and effect, or implication; analogy, including a "known-to-unknown" progression; example; comparison, or contrast; definition, including "expanded" (or explained), and partial or classificatory treatment; temporal or spatial order; simple to complex presentation; "proof," including the inductive and deductive processes; order of importance; and analysis.

But how can this be true if, as our grammar books tell us, the four main types of writing are "exposition, narration, argumentation, and description"? How can the paragraphs that make up writing also be composed in the other ways named, such as by details, example, and so forth?

The answer of course is that the following A types must also belong to the B types. Let us match them up.

A Paragraphs constructed by each of the following *methods*	B usually fall into these main writing *types*
(1) details	(1) description, exposition
(2) (physical) cause and effect, or (logical) implication	(2) exposition, narration, argumentation

[1] Ernest Nagel, *The Structure of Science,* Harcourt Brace and World, New York, N.Y., p. 9 (1961).

A	B
Paragraphs constructed by each of the following *methods*	usually fall into these main writing *types*
(3) temporal or spatial order	(3) narration, exposition, description
(4) analogy, including "known to unknown"	(4) argumentation, description
(5) example	(5) description, narration
(6) comparison, or contrast	(6) description, exposition
(7) definition (including "expanded," partial, classificatory)	(7) exposition
(8) simple to complex	(8) exposition, description
(9) proof (induction and deduction)	(9) argumentation
(10) order of importance	(10) exposition, argument
(11) analysis	(11) exposition

Our 11 ways of constructing paragraphs, then, are *refinements* of the four types of writing.

Method 1 enumerates; it can also involve narration and is excellent for building up clear pictures by appealing to the senses.

Method 2 can be used in reverse to trace an effect back to its cause. In most cases of (logical) implication physical processes are not involved, merely logical relationships. Thus we would not say that a triangle *causes* its interior angles to total 180°, even though they must always do so. Here no causal time sequence is involved.

Method 3 can list events chronologically (time) or can order elements according to position (space). Whether time or space is used will depend on the paragraph subject and one's purpose. Any story told, whether in a report or elsewhere, is chronological. Any object described, as from left to right or top to bottom, is spatial.

Method 4 is vivid and suggestive, but risky, since it never constitutes logical proof. The analogy usually compares (sometimes point by point) something known with something relatively unknown, or at least never thought of in that way before. It is used to indicate similarity of situation or operation and thus to provide new insight.

Method 5 shows how rules apply, and can strengthen an argument. Many writers, and especially speakers, talk too much about principles and generalities and not enough about applications. Myriads of readers and listeners have said to themselves: "What he says sounds all right; but if only he'd give one good example!"

Method 6 divides into the pointing out of similar and dissimilar features. One or the other is chosen depending on what the situation calls for. The perception of similarity, and its opposite, is one of the basic operations of the mind.

Method 7 is most needed at the outset of discussions in order to clarify key terms. To this end, nothing better could be done than to study the nature of definition in a standard logic book. One defines by stating the larger class to which something belongs (its "genus"), then the differentiating characteristics (its "differentiae") that set it apart from any other species of that genus and thus make it unique. Grammatically but fancifully considered, the verb of any sentence definition is like a seesaw equally weighted between its subject (the thing defined) and its predicate (the genus and differentiae definition).

More than a definition can be regarded as an expanded definition. It can also be partial by merely classifying something (stating its genus) but not supplying any differentiae.

The dictionary can be studied with great profit here. How many "seesaw sentences" do you note? Why aren't there more? Can you distinguish genus (plural is "genera") and differentiae in the dictionary definitions?

Method 8 aids comprehension by stating first the things easiest to grasp, then proceeding to the more complicated features. This method of writing paragraphs can overlap many of the others.

Method 9 has various modes, such as the scientific, mathematical, logical, and legal. Here again the two main types of reasoning involved, the inductive and the deductive, should be studied in a logic book. The inductive type jumps from a number of instances or cases to a general conclusion about all such cases, whereas the deductive type starts with the general conclusion and decides that a particular case falls under that type.

Method 10 is almost self-explanatory. One proceeds either from the most to the least important, or vice versa. Newspaper articles (though composed, as we know, of many paragraphs) are written in the former way, so that their reading or printing can be terminated almost at will.

Method 11 is invaluable for comprehension, for it lays the groundwork for much creative thinking and the resynthesis that follows analysis.

The various paragraphs that make up a piece of writing should be instances of the 11 methods of paragraph building and their combinations. Thus, of a passage, the first paragraph might be an example of definition; the second, of details; the third, of analogy; and so on.

Does one have a choice of principles to use for developing a paragraph? Suppose a writer wanted to show that "engineers make good supervisors." After he states this as exactly as he can, using all pertinent conditions, he has a choice of methods. Using illustration and example, he can mention outstanding cases of engineers who became good supervisors. Using comparison, he can explain the capabilities and qualities needed by both. Using cause and effect, he can show that the forces working to make a good engineer—training, experience, and so on—also help to make a good supervisor.

One can see that the use of *different* principles of development extending over several paragraphs would help to clinch an argument.

The opposite thesis, that good engineers do not make good supervisors, would employ the method of contrast.

One may, however, wish to employ a certain principle of development *in exclusion* to all others. For example, in a write-up about forming a chemical compound, the chemist may decide among time order to describe the sequence of operations, details to secure accuracy, or a cause-and-effect treatment to bring to his audience a better understanding of the compound and its components.

Imagine that you want to explain a new automatic feedback circuit to control airplane aileron surfaces. Here are three different ways in which you might present it.

1. *Details*

This feedback circuit consists of a linear potentiometer that converts control surface movement to an electrical signal, a modified wheatstone bridge that compares the potentiometer signal with the original control signal, and an electro-hydraulic actuator that moves the control surface until the two signals coincide. The potentiometer weighs only 2 oz and is made of conductive plastic to reduce the possibility of shock damage. The wheatstone bridge is a modular unit, embedded in dialyll pthalate, which can easily be replaced to introduce various electrical biases into the system. The actuator consists of a piston and cylinder, fed from a pressurized oil reservoir, and controlled by a pair of electrically driven valves that respond to the differential current from the wheatstone bridge.

2. *Cause and Effect*

This feedback circuit assures that the desired control configuration is maintained at all speeds. At supersonic speeds, air pressure on the control surfaces can be measured in tons per sq in. Under this condition, a certain amount of movement of the control yoke will not cause the same control surface movement it would at subsonic speeds. When this (lack of movement at supersonic speeds) occurs, the feedback feature goes into effect, measuring the difference between the control signal and the actual control movement, and then applying additional force to make the two coincide.

3. *Temporal Order*

When the control yoke is moved, two electrical signals are originated. One goes directly to the actuator that moves the control surface, and the other goes through a 3-millisecond delay circuit to an electrical comparator. When the control surface moves it actuates a linear potentiometer, which sends another signal to the comparator. If the two signals do not coincide, the comparator activates a pair of electrically driven valves on the actuator to provide more or less force as required to bring the control surface into the desired position.

Most paragraphs in technical writing will be explanatory, with description and narration as supports. Many paragraphs, in summarizing, will employ proof.

Most paragraphs should have topic sentences and be related to the main subject (which is, itself, often a topic sentence).

Paragraphs can have varying patterns, with the constituent sentences being of various lengths and structures. One paragraph may state the main point, another elaborate on it, others define, analyze, contrast, apply. Each paragraph should have a specific job to perform. A summing-up at the end of several paragraphs is often a great aid to clarity.

From half a dozen books lying at hand, please select about 20 paragraphs and determine their topic sentences, if any, and their principles of development. Such drill will very rapidly inform one about paragraphs. One should ask, is this an interesting and well-organized paragraph? Does the interest consist in its emphasizing something? If so, what? Should the paragraph be cut, rewritten?

TRANSITIONAL PHRASES

Since paragraphs belong to larger patterns of meaning, such as the composition itself (report, article, book, or other unit), individual paragraphs often reveal these larger memberships. They do this by using such transitional phrases as:

furthermore, however, for this reason, in addition, notwithstanding, all things considered, to this end, above all, for example, on the contrary, nevertheless, regardless, but, otherwise, yet, still, meanwhile, presently, finally, in conclusion, for instance, accordingly.

The easiest transitional or linking device, whether within or between paragraphs, is repetition of a key term. Thus the last sentence of a paragraph, "The quality of the *copper wire* was thoroughly checked by this method," might be followed, as the first sentence of the next paragraph, by: "The procurement of this *copper wire* also presented a problem."

Sometimes the transitions or references are "forward looking" ("in the following examples," "we shall now take up in detail," "as will be shown," etc.); sometimes "backward looking" ("as just shown," "in consequence of the last fact," "in other words," etc.). Backward-looking references, especially, help us to strengthen our prose, making the topic of a new paragraph seem to grow from the preceding paragraph. Professional writers make skillful use of connectives.[2]

Entire paragraphs are sometimes "transitional paragraphs," deliberate links in the composition.

The student might here pause again to search for examples of forward and backward references and of entire transitional paragraphs.

"False connectives" should never be used. They try to disguise the fact that the writer has not thought out his subject, has failed to sense the real relationships between the points he is making. Examples are "now let us turn to," and "it is interesting to note in this connection." *What* connection: why not state it?

The previous considerations may be termed "*inter*paragraphic." But the transitional phrases already listed, beginning with "furthermore," occur within paragraphs too, so that they just as well apply to "*intra*paragraphic" transitions. All such phrases perform the function of relating thoughts when needed.

HOW TO REWRITE PARAGRAPHS

To deal practically with paragraphs requires that everything we know about them be combined and applied. We know that they must be subject to the considerations that pertain to sentences, since they are composed of

[2] Please read (Appendix, p. 202) "Atomic Energy—A Fertile Field for Creative Engineers," by Ralph Balent. The italicized words show the transitional and identifying phrases used in this well-organized paper.

them. We know, too, that the paragraph unit has introduced the new considerations of the topic sentence, the principle of development, and the transitional expression. All these considerations are now our concern, to be included as we inspect paragraphs in their native habitat—writing.

Those to follow are chosen (not consecutively) from a privately published article on engineering education in the USSR. To gain practice with written paragraphs, let us see how much general overhauling they would need. Please read this paragraph thoroughly.

1. (a) The following notes are abstracted from a talk given on ——— by Dr. ——— of the ——— Institute of Technology. (b) Dr. ——— was a member of a delegation that visited Moscow, Stalingrad, Stalinov, and Zhdanov to study engineering education. (c) The object of the trip was to assess the quality of technical education in the USSR.

The preceding was the beginning paragraph, the "lead," of the article. Do you not think that it should contain the most interesting fact, theory, or point in the entire piece? After all, engineers and other men of science are people; what is dramatic and newsworthy intrigues them just as it does others. No city editor would accept a lead like this from a reporter. Instead, the who, what, where, when, how, and why (to be included if possible in the first paragraph) should *follow* some interesting lead such as, "Is the Russian system of training engineers superior to ours? Points pro and con are here abstracted from a talk given at. . . ." Incidentally, *where* the talk was given should have been stated but was not.

This first paragraph failed to emphasize the main thoughts (Rule 3).

Sentences b and c both state the object of the trip: "to study engineering education" and "to assess the quality of technical education in the USSR." Need we say this twice, especially in such a short paragraph? If we drop one of the two, which shall it be? The first? Then we would lose the mention of engineering. So let us incorporate that, and use the second. We would have "Dr. ——— was a member of a delegation that visited Moscow, Stalingrad, Stalinov, and Zhdanov to assess the quality of technical education, *especially engineering*, in the USSR." Note the nature of our language: "to" in "to assess" strongly expresses purpose. Therefore it serves adequately to this end, and "the object of the trip was" becomes mere deadwood (Rule 1).

2. (a) The group had difficulty arranging detailed meetings. (b) Evidently the Soviets are not used to meetings having detailed discussion of how operations are carried out. (c) We had a wild-goose chase and failed to arrange meetings due to a clumsy bureaucratic system in which

changes could not be carried out quickly. (d) The Ministry is not the only group involved in education in the USSR.

In this paragraph we find an entirely different set of verbal miscues, some serious, some trifling. From here on, we shall not point out all the rule infractions, but merely comment. We shall mention faults in their order of appearance.

In a, "detailed" in "detailed meetings" does not strike one as the right word (Rule 6). Does he mean "formal," "programmed"? Is he talking about "general discussions"?

In b, a Soviet is a political organization. The speaker simply means "the Russians."

In b, "meetings" and "detailed" are repeated, having appeared in a. If b says something new, as it should, repetition would be wrong as well as unskillful.

In b, "meetings having a detailed discussion" is inept, simply because meetings as such do not talk. If they did, they would have more than the one "detailed discussion" mentioned in the sentence.

Does d have any business being in the paragraph? It does not seem to belong, unless it is expressed differently and with the proper connective. Maybe it should have headed the paragraph, to explain why the group had its difficulties. The sentence should either be dropped, integrated, or used as topic sentence for another, unwritten paragraph. If the sentence were dropped, it would be because it affronts our intelligence: how *could* just the Ministry in a vast nation be "the only group involved in education"?

Note how different the faults in paragraph 3 are from those in 1 and 2.

3. (a) At the end of ten years of elementary and high school, a general examination is given. (b) If a student's grades are high enough, he can apply to technical institutes or universities with expectation of being accepted. (c) Four times as many students want to go to technical schools as are actually admitted. (d) Each student can apply for admission to only one institution. (e) If he is not accepted, he can work a year and apply the following year. (f) Glamor plays a part in the choice of institute and the best students enter the most popular fields. (g) The poorer students must apply to the institutes which have less glamor and teach the more routine subjects. (h) Evidently there is a grapevine system whereby a student can find out the amount of applications for a given institute. (i) If he feels he will not be accepted, he can ask to have his application removed and can then apply at another institution before the deadline.

Does this paragraph strike you as clear? Let us train ourselves to be paragraph diagnosticians by getting a general feeling about the unit before starting alterations. We shall not, however, from here on attempt to be exhaustive in our criticisms.

Does Paragraph 3 lack a topic sentence? Let us try "Getting into a Russian institute of higher education is quite a complicated process"—not the best lead-off sentence in the world, but still much better than the one used.

Is the paragraph still disordered? Then what else is the trouble? Should there perhaps be two more paragraphs here, instead of one? Sentences f and g are about glamor and could be in a separate paragraph. If so, this would move h and i up to follow e: though poorly connected, these two are fairly unified with the rest of the paragraph.

Since getting into school is a temporal process, perhaps a time sequence would help to unify this paragraph. Are any sentences reversed, timewise? Sentence c should probably precede b, since wanting to go to school naturally precedes trying to get in. Sentences e, h, and i also seem to be disordered, since e deals with not being accepted, whereas h and i, which follow, deal with earlier facts. So far, then, just by thinking of time sequence, we see our order should be c, b, h, i, e. We must decide where d would be best placed.

All this bother, you say, just to get something right? *Yes, this bother is the price one pays to achieve, clear, effective writing.* Otherwise what one turns out is like uncooked dough, an abomination to the mind's digestion.

Since b and c are both about admission, we may be able to combine them, thereby shortening the paragraph and increasing the sense of unity (emphasis, Rule 3). Let us try: "Four times as many students seek unsuccessfully to enter technical institutes or universities as those who, with the requisite high marks, are officially accepted."

Reverting to the "glamor" sentences (f and g), we feel that g merely repeats f weakly. Let us clarify this feeling by analyzing them. First let us analyze f: "Glamor plays a part in the choice of institute and the best students enter the most popular fields." This is a compound sentence—composed of two independent clauses—connected by "and." Furthermore, each clause is ambiguous: *what* part does glamor play; isn't it *rather* that the best students *can* enter the most popular fields than that they merely do? There seems to be a cause-and-effect relationship between scholarship and interesting curriculum, which the sentence does not bring out. This intended connection between the two clauses would be clarified if the facts permitted a rewriting such as: "The best students are rewarded by being able to enroll in the most popular fields of study given at the most glamorous institutes." Now, since we have gone to the bother to express the situation correctly, we find that g, being an obvious implication, has become unnecessary.

4. (a) In the study plan for engineers, the humanities consist of Marxism, Leninism, economics of plant management, and history of engineering. (b) The average engineering student considers this to be a good coverage of liberal arts and does not know enough about other things to question this. (c) Students and citizens are proud of the Soviet accomplishments in education. (d) A striking example of changes is the institute at Grunze which is now a center of scientific learning. (e) Forty years ago this area had a nomadic culture.

This short paragraph-in-form-only has three topics instead of one: the study plan for engineers, their opinion of it, and educational advances. If we can combine c legitimately with d and e we shall have secured that "unity" that is clarity's oversoul; a and b are fairly unified. So let us say: "Students and citizens are proud of the Soviet accomplishments in education, which they measure, *not against genuine breadth of outlook,* but by comparison with the earlier crude conditions. A striking example of such progress is the institute at the city of Grunze, which is now a center of scientific learning. Forty years ago this area had a nomadic culture."

Note that the underlined words constitute a *connecting thought,* a supplied transition, which was *implied* but *left out* of the original. Connectives, then, are sometimes implied ideas.

In b, the word "this" occurs twice, referring to different things and both times to entire situations, which is an infraction of good sense (Rule 2, avoid incomplete constructions). The first time "this" means "these courses," the second time it means "this policy." No wonder the sentence is unclear.

5. (a) The best students can take a certain number of elective courses. (b) These are rewards for their achievements and they are always praised for their successes. (c) The halls of the technical institutes have bulletin boards where photographs of the honor students are displayed. (d) One function of the Young Communist League is to prod the laggards and make life miserable for those who do not do well in their studies.

Sentence b seems to be another case like f in paragraph 3—failure to subordinate. A cause-and-effect relationship has been muffed. We can combine b with a to get: "As rewards for their achievements, the best students are praised and can take a certain number of elective courses." The sentence should then be placed in the paragraph that has already discussed the topic, the glamor sentences f and g of paragraph 3.

In b, do the words "these" and "they" refer to the same subject? No Also in b, the word "their" appears twice. Is this what you would call careless, lazy writing?

6. (a) In summary, Dr. ———— pointed out that in the Soviet Union there is a range of quality of education that the student may receive. (b) The most popular and most glamorous technical fields attract the best students and have the best facilities. (c) Less glamorous fields have lower quality students and have poorer facilities.

This was the concluding paragraph of the address. Was it an adequate close for a four-page, single-spaced article containing many colorful facts on engineering education in Russia—especially when you consider that the third sentence repeats the second and that both repeat previous sentences?

Do you think "there is a range of quality of education that the student may receive" is a careful description of what the author found in the Soviet Union? Is this not true of most countries?

An excellent opportunity existed here for a strong summing-up paragraph that would compare, in a memorable way, Russian with American practices in engineering education. The opportunity was rejected, probably because it would require tedious review—about half an hour's work in selecting the strongest and most interesting points of the article.

The rewriting of paragraphs has been discussed by analyzing the faults found in six examples. We saw that all previous knowledge of sentences and paragraphs must be called into use. The first example of the six introduced the concept of a *lead*. The second suggested a new attitude toward a paragraph; namely, to ask whether every sentence of it actually belonged to it. The third example made us see that, if we delete, we then have the problem of rearranging and making consistent that which is left. The fourth example reminded us that transitional and linking phrases belong inside paragraphs, to unify them, as well as between paragraphs, to connect them. The fifth example reaffirmed that analysis and evaluation underly all writing. The last example, the concluding paragraph, pointed up the pressing need to be fair to oneself and one's readers, by giving them the benefit of whatever has been "signaled"—in this case, a summary.

FIVE PARAGRAPHS ANALYZED

Let us begin our 5-5-10 method by discussing five typical paragraphs. The following one was written in 1859.

> When on board H.M.S. Beagle, as naturalist, I was much struck with certain facts in the distribution of organic beings inhabiting South America, and in the geological relations of the present to the past inhabitants of that continent. These facts, as will be seen in the later chapters of this volume, seemed to throw some light on the origin of species—that mystery of mysteries, as it has been called by one of our greatest philosophers. On my return home, it occurred to me, in 1837, that something might perhaps be made out on this question by patiently accumulating and reflecting on all sorts of facts which could possibly have any bearing on it. After five years' work I allowed myself to speculate on the subject, and drew up some short notes; these I enlarged in 1844 into a sketch of the conclusions, which then seemed to me probable: from that period to the present day I have steadily pursued the same object. I hope that I may be excused for entering on these personal details, as I give them to show that I have not been hasty in coming to a decision.[3]

We think the topic sentence is "These facts . . . seemed to throw some light on the origin of species. . . ."

What principle of development was used? Certainly time order, wouldn't you say? Does use of another principle complicate or strengthen the paragraph? He uses details and he is arguing—that is, using proof, though this proof can involve only probability. The social sciences, such as anthropology, cannot, like the exact sciences, avail themselves of exact proofs. We note again the interesting and rather surprising fact that a paragraph can avail itself of several principles of development without being confusing!

What is the status of the transitional phrases? Are there enough of them? Are they properly clarifying? We note many of them: "when on board," "these facts, as will be seen," "on my return home," "after five years' work," "from that period to the present day," "these personal details." Note that all such expressions serve the function of drawing the thoughts of the paragraph together, although they are not the more usual expressions already noted such as "furthermore," "on the contrary," "moreover," "yet," and so forth.

[3] Charles Darwin, *On the Origin of Species,* p. 27, Mentor Books, N.Y. (1958).

As a paragraph, this one seems to be a success. It is introductory in nature and does a good job.

Although the paragraphs of yesteryear tend to be longer, their requisites and general characteristics are no different from those of today.

> The president of a large chemical company was asked, "If a Ph.D. in chemistry came to work for you fresh out of graduate school today and did not keep up with the new knowledge in his field, how long would he be of value to your company?" The company president replied without hesitation, "Not more than three years." The rapid outdating of the chemist's knowledge stems from today's great knowledge explosion."[4]

How did this one strike you? It's good, isn't it? The topic sentence seems clearly to be the last one. What principle of development is used? Look down your list at first, to familiarize yourself with the various means used by a writer. Is it not a clear-cut case of an example?

> The Freedom from Hunger campaign being waged by the Food and Agriculture Organization of the United Nations is also a campaign against thirst. Much of the world's food problem is concentrated in the arid countries, where land, plants, animals and people are parched for water. At the center of the "arid zone" in the lower latitudes is the great desert of the Sahara. Thirteen countries with 148 million inhabitants share this vast territory and the deprivation that its name implies. The Sahara, however, possesses in abundance the remedy for aridity. Below the desert sands in water-bearing rock formations are huge quantities of water to sustain human settlement, pasturage for livestock and, in many places now barren, productive agriculture.[5]

This more technical paragraph, geological in character, is judged by the same criteria as any other. Does it make sense? Does it have a topic sentence? We think it is: "The Sahara . . . possesses in abundance the remedy for aridity." The rest of the paragraph details this statement. What about that part of the paragraph *up to* the topic sentence; does this belong? Although it might have been an introductory paragraph, we think it belongs as a kind of "expanded definition" of the problem of the Sahara.

The principles of development used seem to be analysis and detail.

We cannot see any differences in paragraph making as one ranges from science to science. Here is one from zoology.

[4] E. Noah Gould, "Programmed Instruction—Growing Opportunity for Technical Writers," *Society of Technical Writers and Publishers Review*, p. 9 (April 1966).
[5] Robert P. Ambroggi, "Water Under the Sahara," *Scientific American*, May 1966.

The first sign of change comes when buds near the rear end of the animal's trunk begin to develop into limbs; the jumping legs of the frog. The development of these legs, accompanied by other, less conspicuous changes, takes two to six weeks, depending on the size of the tadpole. In this phase, called prometamorphosis, the animal remains a water-dweller. When the hind legs have grown to about the size of the animal's torso, the tadpole abruptly enters the stage of rapid changes called the metamorphic climax. Forelegs suddenly erupt through small openings in the covering of the gills; the mouth widens and develops powerful jaws and a large tongue; the lungs and skin complete their transformation; nostrils and a mechanism for pumping air develop, and the gills and tail are resorbed by a process of self-digestion and thus disappear. Before the week of climax is over the animal emerges to a new life on land.[6]

How would you set about to judge this paragraph? Developed by time order and many details, it seems unified enough; yet the topic sentence, a kind of summary, does not appear till the end.

Let us take a popular, more "literary" paragraph.

The next question inquired how often people read books that they felt would "advance their knowledge or education in some way." Twenty-two percent replied "frequently," 29 percent said they did so "occasionally," and 46 percent said "rarely" or "never." Another question found that 35 percent had less than twenty-five books in the home, and another 35 percent between twenty-five and 100, with only 27 percent owning more than 100 books (3 percent didn't know). Other questions also were asked about newspaper and magazine reading, and educational achievement.[7]

What is your opinion of this paragraph? Considered in the light of the article's conclusion, it constitutes proof that Americans are not culturally active. Considered internally, it is developed by analysis and detail. When its topic sentence, the first one, is considered, a cause-and-effect relationship is indicated between it and the remainder. So we see again that the fact that more than one paragraphic principle is indicated does not, of itself, make for confused writing.

But do you not agree that a new paragraph should have been started

[6] William Etkin, "How a Tadpole Becomes a Frog," *Scientific American,* May 1966.
[7] Elmo Roper, "How Culturally Active are Americans?," *Saturday Review,* May 14, 1966.

with "Another question. . ."? Do you think the last sentence should have been added?

All these paragraphs have of course been handpicked. You will find, if you examine many paragraphs at random, that most are much more loosely written than the ones we have studied. Too many are paragraphs in name only and too many are made up of slung-together sentences that should be changed or deleted.

The student, to perfect his own style, should read scores of paragraphs and ask himself pertinent questions. Is this a well-organized paragraph? Is it interesting? Does the interest consist in its emphasizing something? If so, what? Is there a sentence or more that does not belong? Should the paragraph have been rewritten?

Here is a heartening thought: astonishing progress in writing can be made by attending especially to the composition of paragraphs.

The following are five paragraphic selections about which the student is to answer the indicated questions. Scratch paper may be used. Answers to this group are in the Appendix, p. 206.

Exercise 22. *Identifying Paragraph Faults*

1. Today virtually every aspect of science is concerned in some way with the atom. Physicians use radiation to treat disease. Mechanical engineers design components for nuclear reactors. Electrical engineers convert the energy of the atom into electricity. Botanists use radioactivity to learn more about plants, and zoologists use it to study animals. Chemists investigate compounds with radioisotopes. Physicists and mathematicians work out the intricate interrelations among the tiny particles of the atom. Agronomists use radioactive materials to improve fertilizers and crops, and nutritionists use them to improve animal diets.

What is your comment on the above paragraph? What is the topic sentence? What principle(s) of organization is used?

2. When at this point one motor stalled, the driver flooded the motor and could not restart it. The service boat went out and towed it in (losing 4 minutes, 3.06 seconds), whereupon it promptly started at the touch of the button. But a lap later, it missed, then stalled again. A mechanic now boarded it in motion and tried an underway check on the fuel supply by switching to an emergency tank. The motor fired smoothly. Water had evidently seeped into the boat's 30-gallon tank.

What is the topic sentence of the above paragraph? What principle(s) of development does it use?

>3. Our general construction is determined by the fact that we are made of living matter, must accordingly be constantly passing a stream of fresh matter and energy through ourselves if we are to live, and must as constantly be guarding against danger if we are not to die. About 5 percent of ourselves consists of a tube with attached chemical factories, for taking in raw materials in the shape of food, and converting it into a form in which it can be absorbed into our real interior. About 2 percent consists in arrangements—windpipe and lungs—for getting oxygen into our system in order to burn the food materials and liberate energy. About 10 percent consists of an arrangement for distributing materials all over the body—the blood and lymph, the tubes which hold them and the pump which drives them. Much less than 5 percent is devoted to dealing with waste materials produced when living substance breaks down in the process of producing energy to keep our machinery going—the kidneys and bladder and, in part, the lungs and skin. Over 40 percent is machinery for moving us about—our muscles; and nearly 26 percent is needed to support us and to give the mechanical leverage for our movements—our skeleton and sinews. A relatively tiny fraction is set apart for giving us information about the outer world—our sense organs. And there is about 3 percent to deal with the difficult business of adjusting our behavior to what is happening around us. This is the task of the ductless glands, the nerves, the spinal cord and the brain; our conscious feeling and thinking is done by a small part of the brain. Less than 1 percent of our bodies is set aside for reproducing the race. The remainder of our bodies is concerned with special functions like protection, carried out by the skin (which is about 7 percent of our bulk) and some of the white blood corpuscles; or temperature regulations, carried out by the sweat glands. And nearly 10 percent of a normal man consists of reserve food stores in the shape of fat.[8]

Why should or why should not this paragraph have been divided into shorter ones? What is its topic sentence? What principle(s) of development do you note?

>4. One more point must be made. The idea that grammar is a logical system has a tendency to make us concentrate on the "concepts" in-

[8] Julian Huxley, "Man as a Relative Being," *Explorations,* Prentice-Hall, Englewood Cliffs, N.J., p. 414 (1956).

volved and to turn away from the study of the actual phenomena of language. A great deal of the material that appears in many texts leads only to the ability to talk *about* the language according to a set of artificial conventions, and has no value whatever in increasing our ability either to use or to understand it.[9]

What is the topic sentence of this paragraph? Its principle(s) of development? Does it employ the principle of proof in any way?

5. Authors of the highest eminence seem to be fully satisfied with the view that each species has been independently created. To my mind it accords better with what we know of the laws impressed on matter by the Creator, that the production and extinction of the past and present inhabitants of the world should have been due to secondary causes, like those determining the birth and death of the individual. When I view all beings not as special creations, but as the lineal descendants of some few beings which lived long before the first bed of the Cambrian system was deposited, they seem to me to become ennobled. Judging from the past, we may safely infer that not one living species will transmit its unaltered likeness to a distant futurity. And of the species now living very few will transmit progeny of any kind to a far distant futurity; for the manner in which all organic beings are grouped shows that the greater number of species in each genus, and all the species in many genera, have left no descendants, but have become utterly extinct. We can so far take a prophetic glance into futurity as to foretell that it will be the common and widely-spread species, belonging to the larger and dominant groups within each class, which will ultimately prevail and procreate new and dominant species. As all the living forms of life are the lineal descendants of those which lived long before the Cambrian epoch, we may feel certain that the ordinary succession by generation has never once been broken, and that no cataclysm has desolated the whole world. Hence we may look forward with some confidence to a secure future of great length. And as natural selection works solely by and for the good of each being, all corporeal and mental endowments will tend to progress towards perfection.[10]

Should this paragraph have been broken up? If so, where? Does it have a topic sentence, a principle(s) of development? To what extent, if at all, does it use proof or "argumentation"?

[9] Louis M. Myers, "Language, Logic, and Grammar," p. 112.
[10] Charles Darwin, *On the Origin of Species,* ibid., p. 449.

SUGGESTED FINAL FOR CHAPTER III

Answers to the next 10 paragraph problems should be written out. Comments on them can then be found in the Appendix, p. 207.

1. For instance, it is necessary to distinguish between "direct" and "indirect objects" in Latin for the simple physical reason that they take different forms. In English the distinction, however fascinating (!) is completely useless. A 4-yr-old can make and understand sentences like "He gave me a book"; and he won't be able to do either a bit better for learning that *me* may be called an "indirect object" and *book* a "direct object." We do not have separate dative and accusative cases in English; and since the boy is not in the least likely to say "He gave I a book" or "He gave a book me," no question of either form or position is involved. If we force him to "distinguish between these constructions" we are not teaching him anything about the use of the language, but only about an unnecessarily complicated linguistic theory.[11]

Would you break this paragraph up for the sake of clarity? What is the topic sentence? The principle of development?

2. And this world state will be sustained by a universal education, organized upon a scale and of a penetration and quality beyond all present experience. The whole race, and not simply classes and peoples, will be *educated*. Most parents will have a technical knowledge of teaching. Quite apart from the duties of parentage, perhaps ten percent or more of the adult population will, at some time or other in their lives, be workers in the world's organization. And education, as the new age will conceive it, will go on throughout life; it will not cease at any particular age. Men and women will simply become self-educators and individual students and teachers as they grow older.[12]

Is this paragraph well integrated, that is, is it obviously all about the same subject? Is it therefore more difficult than usual to determine the topic sentence? What would you say this is? What principle of development seems to be used?

3. All in all, it must be said that intellectual and cultural activity is still distinctly a minority taste. A college education is no guarantee

[11] Myers, op. cit., p. 112.
[12] H. G. Wells, *The Outline of History,* Macmillan, New York, p. 1093 (1924).

of developed cultural or intellectual interests, although it certainly makes such interests more probable. In our rush to get more and more people to college, it should perhaps be kept in mind that half the people who have gotten there show only minor intellectual after-effects. Regarding the other half who can be described as culturally and intellectually involved, the most important question is one that can not be answered by a survey. It is the depth and quality of that involvement. Some years ago I wrote, "There is an urgent need—in fact a national survival need—for invigorating intellectual life, for upgrading the general regard for intellectual excellence. The United States must experience an intellectual renaissance or it will experience defeat. The time cannot be far off—if indeed it is not already here— when the *strength* of a nation, measured in terms of any kind of world competition, will depend less on the number of its bombs than on the number of its learned men." The statement is equally valid today. Unquestionably, there have been changes in recent years in our attitude toward the intellectual life. But the changes have not gone far enough. There is no upsurge of intellectual interest in the young—except in the field of science. Too many people who consider themselves educated have really just gone through the motions. The question that should most concern our educators is not how far they can spread learning but how deep it goes.[13]

The preceding passage was divided into three paragraphs. Try to determine where the breaks came and state whether this was an improvement. How many times is "intellectual" used? Was this justified as an identifying or transitional term? Were synonyms available?

4. The animals other than man live by appearance and memories, and have but little of connected experience; but the human race lives also by art and reasonings. And from memory experience is produced in men; for many memories of the same thing produce finally the capacity for a single experience. Experience is almost identified with science and art, but really science and art come to men *through* experience; for "experience made art" as Polus says, and rightly, "but inexperience, luck." And art arises, when from many notions gained by experience one universal judgment about a class of objects is produced. For to have a judgment that when Callias was ill of this disease, this (medicine) did him good, and similarly in the case of Socrates and in many individual cases, is a matter of experience. But to judge that it

[13] Roper, op. cit.

has done good to all persons of a similar constitution, marked off in one class, when they were ill of this disease, e.g., to phlegmatic or bilious people when burning with fever—this is a matter of art.[14]

What is the topic sentence? What principle(s) of development do you find? Is this a well-integrated paragraph?

5. Every man is rich or poor according to the degree in which he can afford to enjoy the necessaries, conveniencies, and amusements of human life. But after the division of labour has once thoroughly taken place, it is but a very small part of these with which a man's labour can supply him. The far greater part of them he must derive from the labour of other people, and he must be rich or poor according to the quantity of that labour which he can command, or which he can afford to purchase. The value of any commodity, therefore, to the person who possesses it, and who means not to use or consume it himself, but to exchange it for other commodities, is equal to the quantity of labour which it enables him to purchase or command. Labour, therefore, is the real measure of the exchangeable value of all commodities.[15]

Is it not surprising to find "the labour theory of value" announced by Adam Smith himself, "the father of capitalism"? If we are not to regard him as a Communist, how can we resolve this apparent contradiction? Note that when a writer reasons, the quality of his reasoning becomes very important for the integrity of his paragraph. What do you think Adam Smith meant, and did he express it well? Was he mistaken? Please state your view. What is the topic sentence?

6. State clearly the difference between a "true" paragraph and one in form only.

7. Write a brief essay on the topic sentence.

8. What 11 ways of developing the paragraph have been listed?

9. State any original topic sentence and principle of development, then write a paragraph to it of your own devising.

10. Show how a topic sentence of your choice could be treated according to at least two different principles of development.

[14] Aristotle, *Metaphysics*, Book 1, Chapter 1. From Vol. 8, p. B, "The Works of Aristotle," Clarendon Press, 1908.
[15] Adam Smith, *Wealth of Nations*, p. 30, Modern Library (1937).

IV

Turning Out a Written Report

One's greatest help in writing a successful report is detailed knowledge of what supervision requires. To show how technical writing can be thus planned, a Guide Sheet is presented. After a discussion of various report elements, the writer is shown how to avoid the most common faults of reports. The writer's second greatest help in writing is articulate, constructive criticism of a submitted report. To secure this, the use of another form, the Supervisor's Checklist, is suggested. Editorial style is analyzed and described. Discussion of a flow chart of typical photo-offset publishing procedures, a correction sheet for proofreading, and a printshop checklist conclude this orientation in technical report production.

THE PROBABLE SETTING

As the title indicates, this chapter will be more concerned with how to turn out a report than with describing the vast field of reports itself. The ex-student who finds himself employed in a plant or in a large corporation as a technical writer or publisher will not be faced with a bewildering array of report types. The chances are overwhelming that the kind of report he is to concern himself with will be clear-cut and predetermined. It will be taken for granted that his work is to conform to the sort of report already in use. He will not be asked to improve on it or to choose another type, but to follow established usage. Although his firm itself may put out several different types of reports, chances are, again, that he will be asked to specialize in but one of them.

There is another probability. Needs in plants (let us take this word as generally referring to the place of employment) are imperious, but are often filled in a somewhat offhand manner. It is assumed that a person delegated to perform a task can do so. He tends to be assigned to it, rather than trained for it. He may suddenly find it his duty to produce a type of publication that he has never done before. This chapter is directed toward helping such a person.

This chapter, then, is prepared as a general guide to help in producing any sort of report. Discussed are various elements in a report, plus some general observations. The remarks that follow on editorial style are intended to dispel the awe that reports may still command. Closing comments apply to situations in which writer, editor, and publisher are in varying relationships.

A report is a bit of conveyed information. Someone is being told something he did not know. Reports, then, must be myriad—and mostly oral. But those that are to have a certain circulation, and are to endure as matters of record, must be written.

In many firms in which reports are produced, the writer and the editors are distinct persons, working in different departments. The student of technical subjects cannot know in advance if he will be strictly a writer (see "If Writer and Editor are Separate"), or an editor, or both.

Reports originate through executive decisions. For example, consider the engineering report. Engineers have their own supervisors, section heads, and department chiefs, who confer among themselves. Figuring also as originating elements may be sales engineers, "applications" engineers, top-level management, and "customer requests." From an often brilliant and continuous whirlpool of discussion, projects are born and are geared to proposed scientific experiments. Usually one engineer, sometimes several, are placed in charge of doing the needed technical work and writing the report. Before a report reaches the editor for finalizing, it will have been read by and discussed with the engineer's supervisors, and may have been rewritten several times.

PERSPECTIVE OF THE FIELD

The engineering report is only one of a great many kinds of technical reports, yet more has probably been written on the engineering report than on all the rest of technical writing. This is because such a report was the first of the types of technical writing to be "discovered" academically and discussed. But this type probably constitutes a minority of all reports, simply because so many scientific and business activities, other than engineering, require reports. Reports are needed in most of the nonengineering branches of science, in most businesses, in applied economics, in governmental and market research, in military activity, in the law, and in many other quarters.

The best way to describe the various kinds is in terms of their *purposes*. There are sales, operations, research, developmental, marketing, design, product, administrational, and many other types of engineers; distinctive reports are produced from each of these activities.

The foregoing items represent only an *area* classification of the report— the sales area, the marketing area, and so on. A *chronological* type of classification, which divides into period and project subtypes, is also used. Periodic reports can appear weekly, monthly, bimonthly, semiannually, and annually. The well-known project report divides into preliminary reports; progress, or interim, reports appearing irregularly; and final, or completion, reports that sum up entire projects, including their last stages.

A report can also be named according to the *source* of its information, as for example a *laboratory* report. Considered technically, furnaces, engines, test-stands with their mazes of hydraulic or other tubes, oscilloscopes, and many other kinds of appliances are parts of laboratories. The lab report, increasingly important in an era of industrial research, bristles with statistics and requires apparatus and process description and skilled interpretation.

The more one learns about reports, the more difficult it becomes to classify them neatly. For example, the "proposal" is very important in modern industry. It is a special kind of report, often based on data supplied by engineers and stating under what conditions, including price, a manufacturer stands ready to supply a product or service. It is a deluxe effort to sell the company's offering, usually prepared with scrupulous care by professional writers and editors.

The prominent "R and D" (research and development) report, intended to make discoveries known, is partly in the domain of pure science; yet it can cause production lines to work overtime.

The "balanced," or "normative," report equates stubborn facts with an account of changes needed in order to meet requirements or achieve ideal situations. That is, it balances a report of the actual plant situation with a statement of the required remedies.

Design summary reports, topical reports, technical data reports, memos— the list seems endless. Each type serves its proper purposes. For instance, in one concern the memo means "a working paper which may be expanded, modified, or withdrawn at any time, and is intended for internal use only. Further dissemination is not permitted, and distribution to abstracting agencies is not authorized."

Reports may be classified in so many different ways that a single one can belong to several classes or cross-divisions. Thus an R and D report, for example, can also be a lab and progress report.

Such considerations as the physical appearance of a report, that is, whether it is "formal" or "informal," long or short, and so forth, though much dwelt upon, are superficial.

We repeat, the best way to describe the various types of reports is in terms of their purposes. This means that a technical writer must identify and keep in mind his target public and describe and explain what is meaningful and pertinent for them. Description of a new speedboat, for example, would be written from one point of view for the professional racer, enthusiastic about speed and design; from another for the mechanic who will service and repair it; and from still another for the manufacturer who plans to build in quantity. This "slanting" for the specific target public is usually very helpful in narrowing and simplifying one's task.

THE RESEARCH ACTIVITY

The extent to which research need be done can vary greatly from one report to another. Sometimes writing is done entirely from laboratory data, no research work being necessary. Other reports cannot be made without using knowledge to be found in various technical volumes.

Generally speaking, the best single thing an author who needs help with research can do is to inform a reference librarian, in a large city or university library, of his needs. The access that such authorities have to special knowledge is tremendous; they can tap the latest, most trustworthy sources.

Thousands of practical, well-indexed reference books have been published in the last two or three years. They include such subjects as automatic controls, computers, construction, drawing, electricity, electronics, fluid mechanics, gears, human engineering, industrial management, mechanics and design, metals, nonmetals, nucleonics, production, strength of materials, thermodynamics and heat transfer, transport and engines, and vibration.

Also available are many excellent *standard* reference books, often called handbooks. A few titles are *American Institute of Physics Handbook*, 1535 pp.; *Materials Handbook*, 950 pp.; *Handbook of Engineering Fundamentals*, 1322 pp.; *Electronic Engineer's Reference Book*, 1588 pp.; *Standard Handbook for Electrical Engineers*, 2248 pp.; *Machinery's Handbook*, 2104 pp.; *SAE Handbook* (annual); *Civil Engineering Handbook*, 1184 pp.

Van Nostrand's *Scientific Encyclopedia*, the best general reference book on scientific subjects, is issued annually in one huge volume. The *Dictionary of Science and Technology* (formerly *Chambers*), edited now by T. C. Collocott, is also standard. Equally good is the *McGraw-Hill Dictionary of*

Scientific and Technical Terms, edited by Daniel N. Lapedes.

Further aids in discovering basic material for one's reports are the bibliographies. They include *Scientific and Technical Information Sources,* by Ching-Chih Chen, M.I.T. Press, 1977; *Guide to Basic Information Sources in Engineering,* by Ellis Mount, published by John Wiley, N.Y., 1976; and *Science and Engineering Literature: a Guide to Reference Sources,* 2nd ed., by H. R. Malinowsky, published by Libraries Unlimited, Littleton, Colo., 1976.

An R. R. Bowker Co. annual, *Books in Print,* lists 110,000 books under 24,000 subject headings. The Special Libraries Association publishes, in ten issues a year, a *Technical Book Review Index* and puts out the succinct *Bibliography of Engineering Abstracting Services.*

Abstracts give summaries of articles or books, rather than merely indexing or listing their titles. The best known of these is the biweekly *Chemical Abstracts,* which supplies data from 7000 periodicals. Other abstracts are: *Technical Book Review Index, Science Abstracts, Applied Mechanics Reviews, Engineering Index, Applied Science and Technology Index,* and *Business Periodicals Index.* Writers should learn what publications exist for their special fields.

Two abstracting and indexing services in particular cover the technical literature extensively. The *Engineering Index* abstracts domestic and foreign articles from 1200 periodicals and 300 books annually. The abstracts, printed on small cards, are mailed weekly to subscribers. The 300 subjects covered may be subscribed to individually or in groups. If one takes the complete service, he receives a total of about 25,000 cards a year. The second comparable service split in 1958 into the *Applied Science and Technology Index* (which covers 199 science, engineering, and industrial publications) and the *Business Periodicals Index* (covering 120 periodicals in business and finance).

You might even be entrusted with the greatest of all researching responsibilities—forming a company library. Two helpful books are: *Scientific and Technical Libraries; Their Organization and Administration,* by L. J. Strauss, published by Becker, N.Y., and *University Science and Engineering Libraries; Their Operation, Collections, and Facilities,* by Ellis Mount, published by Greenwood Press, Westport, Conn. Also available are: *How To Use the Business Library,* by H. W. Johnson, by Southwestern Publishing Co., Cincinnati, and *Business Information Sources,* by Lorne Daniels, University of California Press.

Two reliable tests of when one is ready to begin writing his report are (1) when he can write a limiting sentence, or name a topic that, colleagues agree, is the theme of his report, and (2) when he can transfer his card outline to one sheet of paper. He may then throw away his cards, which have served their purpose. The writer should be sure to examine this one-page outline with a gimlet eye. Are there weak spots, or steps left out entirely? This once-over is the last chance he has to strengthen the outline.

THE GUIDE SHEET

One of the causes of poor reports is the lack of professional communication between the writer and his commissioning supervisor. By "professional" we refer to the profession of writing; the supervisor does not talk technically enough about what he wants and expects from the writer. By "commissioning" we mean that superior who gave the writer his task and to whom the writer is responsible. The supervisor talks about the subject of the report, but not about *how*, specifically, the subject should be treated. He may seem to be giving the writer a great and generous leeway. But when the writer has done his work and shows it, the supervisor says, "Oh, but you didn't put *this* in," or "You shouldn't have said *that*," or "I'm sorry if I gave you that impression." If the writing job had been talked about adequately beforehand, chances are those faults would not have occurred. As it is, the supervisor is displeased and dissatisfied, and the writer is justifiably embittered at the lack of communication between himself and his superior.

So the "guide sheet" has been drawn up for use in all such commissioning situations. The oblong spaces at the top are to be titled with the names of the various types of writing used in each concern. These names vary almost as much as do those of racehorses. No two concerns—even subdivisions of the same corporation—will name their writing products similarly, even when they are similar in nature.

The 16 items on the guide sheet prompt the supervisor to explain to the writer how he thinks the report should be written. These items are not final; they all simply represent the *idea* of working up an adequately detailed guide sheet so writers will not receive scanty instructions. The items are to be reworked until they exactly fit the needs of each concern. Then, when thoroughly conforming to the reports of the individual firm, the form will be a valuable aid in writing. See p. 109.

POSSIBLE ELEMENTS IN A REPORT

We want to warn the reader that he may not receive any help in noting and determining just what elements compose a company report where he works. For formal understanding, he should rely on his own analysis of the situation.

The author's writing of the basic report elements requires care, since these are both numerous and complicated. There need be no general agree-

GUIDE SHEET FOR WRITING ASSIGNMENT

TDR*☐ MEMO ☐ TOPICAL ☐ PROPOSAL ☐
BIMONTHLY ☐ SC. ARTICLE ☐ BROCHURE ☐ OTHER☐
Title_____No._____
Author_____Reviewer_____Date_____

Supervisors or other reviewers should discuss with assigned authors, before rough draft is started, every pertinent heading below, in order to minimize rewriting.

1. SPECIAL FORMAT REQUIRED, OR SPECIAL REASON FOR WRITING:
2. ESTIMATED NO. OF PAGES:
3. TIME AND SERVICES AVAILABLE (before editing and printing):
4. SCOPE (limiting of subject):
5. SLANT (intended audience):
6. DRAFT OF ABSTRACT:
7. INTRODUCTION (in this case, what should it include?):
8. ADVISABLE OPENING TO AROUSE INTEREST:
9. EMPHASIS TO BE PLACED ON:
10. SUBJECTS OR POINTS OF VIEW TO MINIMIZE:
11. NO. OF DISPLAYS NEEDED: a) Line drawings—
                               b) Photographs—
                               c) Tables—
                               d) Sketches—

12. RESEARCH ADVISED: a) History of project—
                         b) Company policy—
                         c) Promising line of thought—
                         d) Other company areas involved
                         e) Literature available; e.g., System Description, TDR, etc.—

13. SUGGESTIONS FOR DEVELOPMENT:
    (Often made after work has
    revealed best line of attack)
14. MATHEMATICS NEEDED: a) Type—
                           b) Amount—
15. ARE THESE ELEMENTS REQUIRED?   a) Results—
    (Refer them clearly to   b) Conclusions—
    purpose of writing)   c) Recommendations—
16. OTHER REMARKS:

*Technical data report

ment about the report's elements because the thousands of report-issuing bodies use many different sets of elements, depending on their needs. Although very few concerns use them all, the following are the major ones.

1. *Cover*

Used on a "formal" report only, as a separate page, the cover shows the report title; usually, company identification marks; security classification if any; and often, the number of each copy. Warning: Choosing a cover can involve not only selection of paper or other stock, but also type of binding to be used.

2. *Title Page*

This page shows the title again; usually the names of the authors; perhaps approval signatures and date; again, any security classification; and number of contract. The title should be (a) as brief as possible without sacrificing meaning, (b) descriptive, but not in a sentence form, and (c) without articles ("a," "the") or needless punctuation. If the report has been revised, the revision date should appear at the bottom of the page.

3. *Letter of Transmittal*

This is a letter attached to the report and addressed to the original requesting concern or authority. Sometimes it substitutes for the preface, being then bound into the report. Where reports are distributed internally (inside the company) and regularly, or externally but regularly to an official list of perhaps hundreds of firms, the letter of transmittal has no place. But whenever a special report has been particularly requested, great care should be taken to deliver it to the requester personally. The author(s) also should be fully briefed on their own report so they can not only summarize it in an oral presentation, but can answer questions. Specially requested reports should not be delivered in an offhand manner, but deliberately and almost ceremoniously. This pleases the requester, being more personal than the letter of transmittal, whose place the personal delivery may take.

4. *Table of Contents*

This is the outline of the report, being a list of the first-order headings, and perhaps of the second- and third-order headings. The exact format adopted here is a matter of company style.

5. *List of Tables*

This consists of a list of the formal (i.e., numbered and titled) tables in the report.

6. *List of Figures*

These are simply the illustrations used in the report, consisting of line drawings, photographs, charts, graphs, diagrams, and so forth. This is the "art," often essential for clarity or speed of comprehension. Reports may, however, need no figures. (Elements 5 and 6 may be regarded as part of 4.)

7. *Preface or Foreword*

The function of this element is to present, briefly, official or administrative facts about the report, such as what project initiated it, and under what explicit authority it was conducted. Especially helpful persons may be mentioned here, instead of in an "acknowledgment" item.

8. *Abstract*

This is a technical summary of the entire report, written ideally in a single paragraph of about 150 words, so that librarians can place it on small cards. It usually states, precisely and briefly, the purpose of the report, work done, and conclusions reached. The greatest value of the abstract is to those who wish to read no further. (Sometimes when lengthened to cover a long report, it is termed a summary.) Thus the abstract is used (a) as a bibliographical device by librarians, and (b) to satisfy special interest. Word for word, it is probably the most difficult part of the report to write properly.

The published *Technical Notes* of the National Aeronautics and Space Administration include two pages at the rear containing eight index cards for librarians to place in file drawers. On each of the eight is printed the abstract of the Technical Note.

One such reads as follows:

> With ionization and nitric oxide formation neglected, analytic expressions are derived for the composition and thermodynamic properties of nitrogen-oxygen mixtures. The assumption of negligible nitric oxide is later removed and solution by iteration is used to prepare composition tables for the chemical species N, O, N_2, O_2, and NO. For densities sufficiently high for ionization to be negligible, the results of the study are applicable to 10,000°K.

This succinct abstract runs to only 68 words. Title, author, and date precede it.

An abstract, then, does two things: it provides a library indexer with appropriate categories for proper filing, and it tells a literature-searching person the value of the report to him. If an indexer cannot classify an abstract properly, the literature searcher will not find it in the category to which it belongs; and if it is not informative, it will be nearly worthless to him. An abstract should be a very condensed account of why work was done, what was done, and the conclusions. If the report belongs to applied science, a statement is valuable as to how the activity written about is reducible to practice.

Some authors prefer to write an abstract before they start the report because the abstract's careful composition will help to organize the report. Then if the report's development leads to reevaluation of major items, the abstract may be amended.

With so many articles competing to be read today, the average scientist keeps abreast of his field by scanning the abstract journals; he will be most apt to read those articles whose abstracts are not only very clear but communicate a sense of urgency. Abstracts should also appeal to non-specialists: one test of whether an abstract is clear is to see if those with degrees in other fields can state its problem, its approach, and its summary.

Both title and abstract of a report should be as well written as possible, since they appear not only on library index cards but in abstract journals and in publication lists. Titles longer than eight words should be suspect; phrases such as "some aspects of" or "some characteristics of . . ." can be deleted.

An abstract can sometimes be formed by making a significant sentence out of each first-order heading, the collection of these most-important statements being the abstract. This demands of the writer the ability to separate unerringly ideas of first importance from all the others.

Repetition of ideas is particularly wasteful in an abstract. What is the repeated idea in the following abstract of an article on "titanium carbide materials for high-temperature applications"?

Titanium carbide, when cemented with a suitable binder, is an effective cutting material, being tough, hard, and having high resistance to abrasion, corrosion, heat, and oxidation. And it is cheaper than tungsten carbide cermets. Aside from use as a cutting tool material, titanium carbide has been considered as a structural metal for high temperature use. Several binders for this purpose have been proposed. Density and strength are needed for use in turbine blades.

The repeated idea is "aside from use as a cutting tool material." Also, ideas of less than primary importance have been included in this abstract; mere studies and proposals seem secondary. One feels the abstract should be much shorter, something like the following:

> Titanium carbide, when cemented with a suitable binder, is an effective cutting material, being tough, hard, and having high resistance to abrasion, corrosion, heat, and oxidation. Cheaper than tungsten carbide cermets, titanium carbide provides density and strength for use in turbine blades and other structural elements.

9. Introduction

This part of the report includes several subparts, to be used as varying conditions indicate.

One begins the introduction by stating the problem with which the report deals and his consequent specific purpose.

This problem/purpose is necessarily related to a report background, which is therefore stated if it adds to the reader's understanding and appreciation of the subject. Its status up to the present can thus be given, after perhaps sketching the history and relevance of prior studies.

The writer's unique contribution, if any, should then be stated, including noteworthy conditions under which the work was done. The ranges of variables considered, the types of tests and analyses made, may be given, although not—in this relatively brief section—in detail.

Here is the place to mention, if advisable, what the report does *not* intend to accomplish, as well as its necessary omissions and limitations. In these days the report writer is especially obligated to be honest because he is under so much stress of time, compared with yesterday's investigators, that his report may be unavoidably incomplete. He may have discovered in the course of his investigations that certain crucial tests are still to be made; things may even have gone wrong. In any event, authors should not hide the serious faults in their own work, or use words as a smoke screen to obscure the correct view of a situation. A report is a moral as well as a technical product.

10. Body

This section tells in detail what was done, and how. It may include much careful description and explanation. We say "careful" because the author is writing to his target public, not to his close colleagues.

The body may prove conclusions already stated in the abstract and introduction (leaving *detailed* proof for the appendix).

Note that the term "body" is the only one in the list of report elements that is not generally repeated as a heading in the text of the report itself. One thinks nothing of using such terms as "introduction," "results," "conclusions"; but he leaves out "body" in favor of specific headings. In the following outline on a radioactive element, what makes up the body?

    Abstract
   I. Introduction
  II. Experimental Procedure
 III. Results and Discussion
     A. Tensile Properties
     B. Creep Properties
     C. Thermal Cycling
     D. Metallographic Studies
     E. Hardness
     F. Chemistry
 IV. Postirradiation Evaluation
  V. Conclusions and Recommendations

Would not all of II, III, and IV consitute the body of this report?

The names of report elements other than "body" may also disappear. Here, for instance, are the first-order headings of a Department of Engineering, University of California report entitled "The Air Pollution Problem in Los Angeles."

    I. Introduction
   II. Air Resource Test Facility
  III. Microwave Methods of Studying Air Pollution
  IV. Pollutant Prediction
   V. The Problem of Incineration
  VI. Automobile Exhausts
 VII. Eye Irritants
VIII. Removal of Irritants from Polluted Air

The only report element we recognize is "introduction." Yet the report was well organized.

Except for the abstract, the hardest part of a report to write may well be the body. First-order headings composing the body might refer to such matters as equipment used; procedures; relevant tests; location; dimensions; physical properties; appearance; how data were processed and interpreted; principles of construction used; analysis of methods; alternative solutions; mechanical, chemical, metallurgical, and electrical characteristics; discussion of experimental procedures; investigation of materials or constituent parts; sources of information.

11. *Results*

Here are presented factually and impersonally the basic findings of the report, whether favorable or unfavorable. Since these data can be as raw as the mere readings on instruments, they often lead to the next element.

12. *Conclusions*

These are inferences that are based on and interpret the results, therefore constituting opinion lodged in fact. The conclusions should provide an answer for the problems originally stated, but need go no further. Results and conclusions naturally work together (although sometimes management wants only the results, wishing to form its own conclusions). To find that a machine produced 300 items of a certain product per hour would be a result. Using knowledge and experience to decide that the machine is worth the price asked for it is a conclusion. Of course a conclusion may be more general than this; it can state whether—and if so, to what extent—scientific knowledge has been advanced by the report.

13. *Recommendations*

These are the author's suggestions for action. It is a matter of company policy as to whether reports include them. Continuing the case mentioned, one could recommend that his concern buy the machine and put it to immediate use.

14. *Appendixes*

These are relatively minor parts of the report (referring usually to the body of it) and are best placed before the reference list. They are comprised of material that is needed to fully understand the report but disrupts its continuity if not detached. An appendix might explain how equations were derived, or give a specialized experimental procedure not published elsewhere, or a detailed description of equipment used, and so forth.

15. *References*

These usually appear, collected, at the end of the report. They help readers who wish to study further or to check statements made. The choice of marks used to refer from text to reference page is a matter of style. The term "bibliography" refers to books and articles that will generally increase the reader's knowledge of the report subject.

Many, if not most, reports omit the letter of transmittal, the preface, and one or two of these: results, conclusions, and recommendations.

GENERAL OBSERVATIONS ON REPORT WRITING

There are so many different kinds of reports, and each report-issuing concern is so set in its own ways, that helpful generalizations are hard to make. Those that follow here and in the next section may be used as guides when applicable.

Technical reports that concern science or engineering usually involve either scientific *objects* or scientific *processes*. A layout for a proposed machine shop, or for a cloverleaf highway intersection, a setup for a certain research project, analysis of a deposit on certain furnace coils, track maintenance on railroad Z—these involve physical objects rather than processes. But such matters as how to replace a broken windowpane, how to measure coil resistance with a wheatstone bridge, the causes of heat losses in an internal combustion engine, how to use the slide rule, fractional distillation of crude petroleum, the miniaturizing of transistor radios—these concern processes more directly than they do the physical objects associated with them.

It is important to distinguish scientific objects from scientific processes because they should be given different treatment in the body of a report.

True, the first step is the same for both: to expand the definition. Both should be defined as precisely as possible, then described by adding characteristics through the use of analogy or example; in brief, by adding to the definition whatever will make the object or process stand out unmistakably in the reader's mind.

But the second step is unique for each. The overall construction of the scientific object should be presented, followed by an explanation of each of its major parts. In the case of the scientific process, the materials, tools, and conditions needed for its operation should be stated.

Step three for the scientific object is to tell how it works, including the principles involved; for the scientific process, a narration of each major step in the process is called for.

Step four is alike for both: an evaluation of the object or of the process. Depending, then, on which is involved, the manner of treatment is as follows.

Scientific Object	*Scientific Process*
(1) Expanded definition	(1) Expanded definition
(2) Overall construction and construction by parts	(2) Naming of materials, tools, conditions required
(3) Statement of how it works, including the principles involved	(3) Narration of each step or stage in the process
(4) Evaluation	(4) Evaluation

Even this skeleton outline of how to handle the body of a report may be modified. For example, sometimes an evaluation will not be desired, or the subject of a report will be as much about a scientific object as it is about a scientific process. Consider the budgeting operations of an aircraft maintenance shop. Budgeting is a process, yet without the physical object of the shop itself, nothing could be done. "Track Maintenance on Railroad Z" would be another such hybrid.

A still simpler rule of thumb for writing the body of a report is to answer these three questions: (1) What is it? (2) How is it assembled? (3) How does it operate? If this routine is not applicable, there should be no effort to *make* it apply. But if it does apply, one need ask only if Step 4, an evaluation, is wanted.

HOW TO AVOID THE MOST COMMON FAULTS IN REPORTS

1. *Clearly State the Purpose*

Authors frequently assume that the purposes of their reports are known because they and their colleagues know them. The target public, which may be thousands of miles and some months away, is temporarily forgotten. To check on this situation, the author should state the purpose of his report clearly to himself—then see if he has said as much in the manuscript itself!

Repetition, the greatest of the laws of learning, can be applied here. Why not restate the purpose once or twice in the course of the report? Readers, relatively unfamiliar as they are with the material, might be grateful.

We have noticed how some persons, in answering a question, talk all around it instead. It is possible to talk all around the purpose of a report. So let's make sure to state this purpose clearly.

2. *Give Brief Background of Report Topic*

There is something about knowing the background of a project that stimulates interest. An entire subject can come to life because its background is supplied. For example, food-packaging may be a dull topic until it vitally concerns the well-being of astronauts. Computer-produced writing can gain greatly in appeal when we are told it will involve raised surfaces for the use of the blind.

Every subject, of course, has a background; otherwise it would not fit into the scheme of things. We can easily form the valuable habit of deliberately envisaging a subject's background, the more clearly to present it. The place to do so is ordinarily the introduction of a report.

A background is to be described or stated only when it will enhance the report. Interim reports—those put out at short intervals on a continuing subject—usually do not include the background, since it would have to be repeated again and again.

3. *Work from a Strong, Clear Outline*

The penalty for not doing this is to produce a report that does not proceed from one carefully connected, obviously related part to another. An outline need not obtrude, any more than the skeleton of a body does; but it should hold up the entire body and be felt beneath it, so to speak. The payoff for a thoughtfully prepared outline is a complete, logically organized report.

Think of building anything, in particular a house. One always starts with the foundation. Otherwise, the predictable result is some jerry-built edifice that will not stand up and last as it should. The analogue in writing is a report that seems to lack structure and thus gives the impression of being formless and rambling.

It was the aim of Chapter I to firmly establish outlining as the correct preparation for report writing; it would serve as a good refresher for point 3.

4. *Use Enough Visual Aids*

Sometimes a reader craves a picture of what is being talked about. Sometimes he pines for a graph that will show him at a single educated glance the complex relationships over which the text is laboring. The function of technical art is to supplement the written material, make it more easily understood. When art will do this, it should be used; for example, a graph is usually much more clearly grasped than the table it expresses. The graph, with its dependent variable(s) plotted against the independent variable, is simple; the table, with its many entries, its rows and columns filled with varying quantities, is complex. A graph can be a great relief to the reader.

In order to become more resourceful in using visual aids in reports, the writer can study the major types such as line drawings, the halftones, block diagrams, and the flow charts. Any artist in a company's employ should be able to guide a writer, as he is familiar with the various types even though he may specialize in only one or two of them. By studying these types, a writer or editor may realize that artistic possibilities for a manuscript are more extensive than he had at first thought. He can readily see that a thing need not yet exist and be photographable in order for it to be representable in an artistic visualization. An artist can give valuable advice as to how he may be able to clarify and add interest to the writing.

5. *Arrange if Possible for Pleasing Layouts*

The appearance of a report that is not well laid out can be annoying. It is sometimes difficult to know why, so important in this matter are our subconscious impressions. Unexpressed psychological preferences are often involved. The subject is discussed under "Editorial Style" in "1. Format," p. 127.

See if the way a report is "gotten up" is attractive; if not, whether the reasons can be put into words. With a little effort they usually can. Are the margins too narrow? Should the lines of type be spaced farther apart? Is the ink faint, uneven, or the wrong color? Half a dozen considerations can be involved.

A good "layout man" or person with experience should diagnose these situations. Maybe the copy is not broken up enough with subheads. The different kinds of type used for body type and headlines may not look well together. There *must* be a reason or reasons for the unpleasant impression, since there is no effect without some cause.

It may be impossible for the writer or editor to change certain unsatisfactory conditions of layout. Typewriter type, for example, cannot be changed at will; but such things as margins can. The unpleasing condition, once identified, may be one of those that are alterable. It is the duty of those involved to think about the matter.

6. *Be Sure Conclusions Are Supplied, if Desired*

It is often said that the facts speak for themselves, but this is false. Facts are mute; they must be noticed, grasped, and have inferences drawn from them. Facts are just inert dough until alert minds appreciate their significance and act accordingly. Conclusion-drawing, *when* conclusions are expected, should be done with clarity and proficiency.

Scientists and technical men can be strong in experimentation but weak in drawing inferences or conclusions. Conclusions are easy to draw in some experiments, as when one learns that airplanes flying beyond Mach 1 or the speed of sound do not disintegrate. From this one can easily take the next step and conclude that airplanes can safely be built for flights beyond this speed. That's an easy one.

But what if an experiment involved the testing of cladding materials for nuclear fuels—and the materials proved inadequate? A conclusion not quite so easy to make would be to test other materials.

Or what if another material proved inadequate for another purpose? It might still be excellent for making a new product entirely, such as a super-toy. It takes real application, however, to pull valid conclusions out of the air.

7. *Make Language Precise and Concise, not Vague and Verbose*

This is a writing task that must be worked at constantly. The diction of some reports is a sad blend of vagueness and verbosity. If the right word is not thought of and the wrong word is used instead (thus assuring vagueness), words that do not belong are shanghaied into the sentence in a depressing attempt to make up for the lack of vocabulary.

The best preparation for applying this rule is to constantly study vocabulary so that "specific" words—those that correctly express one's meaning— are at one's command. To the extent that they are, we need not think up added words to approximate our meaning.

ACTIVE OR PASSIVE VOICE?

A well-known language problem in writing reports is whether to favor the active or the passive voice. Examples of the active voice would be "Bob threw the ball" and "The surveyor measured the slope distance." Examples of the passive voice would be "The ball was thrown by Bob," and "The slope distance was measured by the surveyor."

Thus to change a sentence from active to passive voice we take the object ("slope distance") of the verb ("measured"), in the active-voice sentence, and make it the subject of the sentence in the passive voice. Also, in this case the auxiliary verb "was" is placed before "measured"; the preposition "by" is used to make a prepositional phrase ("by the surveyor") of the former sentence-subject ("the surveyor").

If the prepositional phrases themselves are dropped, sentences in the passive voice can become vague indeed, retaining the action but not the agent. For example, "The ball was thrown ~~by Bob~~" or "The slope was measured ~~by the surveyor~~."

Suppose one must narrate a hundred or more sentences about a process; shall he use the active or the passive voice?

The answer depends on whether one wishes to feature the agent or the action, the operator or what is operated on. The passive voice is often preferred in cases in which the operator is a physical but lifeless agency. Otherwise we should be treating parts of a process as though they were agents, saying for example, "The steam coils melted the sulphur" or "The cotton linters, so treated, manufactured the smokeless powder," instead of saying "The sulphur was melted by the steam coils" and "The smokeless powder was made from cotton linters, treated. . . ."

Sometimes such lifeless materials can be so described, without strangeness, as "The exhausted water solution produced 2 percent ammonia by weight," or "The wheatstone bridge measured the electrical resistance of the coil." The only conclusion here is that judgment must be used as to whether lifeless subjects are to be written up in the active voice. In these two cases—and there are many others—water and a bridge are referred to acceptably as though they were bona fide agents.

But this choice between using active or passive voices involves other difficulties. There are situations in which no actor needs to be specified. The writer could presumably state who conducted the experiment or who filed the report, but there is no point in his doing so. He can say "The experiment was conducted" or "The report was filed" because there is no interest in who did these things. Use of the passive voice is justified in such cases. Technical writers are often confirmed users of the passive voice because they wish to emphasize the what and the how rather than the who. Unfortunately, however, this can be done so as to show what is called "vacated responsibility." The writer says "it is proposed," "it was inferred that," "it has therefore been decided," until the reader has no clear idea of who did what or of who is responsible for anything in the report. Unwise use of the passive voice therefore can cloak a report in an annoying anonymity and can give readers as false an impression of authors as does the overuse of "I."

When it is wise to avoid continual personal references, and when an author should draw attention to an object or process rather than to the agent or to the cause involved, use of the passive voice is justified. Otherwise the active voice is preferable.

THE SUPERVISOR'S CHECK LIST

The guide sheet, as we have seen, is to be filled out and discussed fully by supervisor and writer *before* any writing is done, whereas the checklist is to be used by the supervisor *after* the writing is done. Nearly all supervisors must edit, although few have training for the job or a knack for editing. The checklist device is designed to put words into the supervisor's mouth. He is probably not a writer; he just "knows what he wants." But he may not quite know how to say it—hence the checklist.

On pp. 123–124 are 37 faults of various types that may be found in a manuscript. To use the checklist, the supervisor merely writes down, in the left-hand margin of the manuscript, at the place in question, the number of whichever fault is involved. The author, receiving the marked-up manuscript back, refers to his own copy of the checklist and knows that where he reads, for example, "4," his supervisor wants what is written there to be discarded as irrelevant. A "28" will mean that the punctuation should be improved, and so on.

This checklist is more apt to be usable without changes than is the guide sheet, simply because certain writing faults are more universal than are the various considerations that might be pertinent to writing a report. Nevertheless, the Checklist too is an *idea* that should be adapted for use according to the specific needs of each concern. For example, if there is no mathematics in its reports, there would be no need for Item 19.

After the writer receives his report back from the supervisor, he will want to discuss the indicated criticisms directly with him. They now have a basis for discussion whereas, without such a form, the supervisor may know "something" is wrong with the writing, but not what!

To repeat, the guide sheet and the checklist should be revised to meet exactly the needs of each report-using concern. Then both may be printed up and used as forms by those individuals who commission writers. The forms are designed to lubricate the sometimes sticky and difficult relationship between supervisor and writer and to facilitate good writing; they are designed not only to save talking and mistake-making, but to create more fruitful talk about writing.

SUPERVISOR'S CHECKLIST

TITLE_____NO._____

REVIEWER_____AUTHOR_____DATE_____

(a) Encircle faulty words, lines, or passages on manuscripts themselves.

(b) Write number of each fault, as listed below, in left-hand margin opposite relevant passage. Add comments freely.

GENERAL FAULTS

1. State or clarify your purpose.
2. Emphasize and/or add to this part.
3. Too broad a statement; qualify it.
4. Discard as irrelevant.
5. Rearrange into logical or chronological order.
6. Too vague. Add explanatory details, etc.
7. This consideration seems misplaced.
8. Change the tone———mood———tense———voice.
   a. Use active voice.   b. Use passive voice.
   c. Use imperative for directions.   d. Other change.
9. Take a definite stand or draw a conclusion, even if negative.
10. State clearly the reasons for your conclusion.
11. Strengthen the argument; make it valid.
12. Confer on question of policy.
13. Other

SPECIFIC RECOMMENDATIONS

14. Make an outline, and rewrite from it.
15. Write abstract. Make it shorter———clearer———longer.
16. Define the underlined topic in the text.
17. See that headings in table of contents are identical with headings in manuscript.
18. Is this the correct terminology?
19. Clarify the mathematical derivation.
20. Reference data properly———add a reference———repeat the reference.
21. Change Section———into an Appendix.
22. Add———delete: results———conclusions———recommendations.
23. Other.

SENTENCE FAULTS

24. Break up into shorter sentences.
25. Delete words; state more directly.

26. Change word order.
27. Express more concretely.
28. Improve punctuation.
29. Rewrite along these lines.

PARAGRAPH FAULTS

30. Change sentence order.
31. Begin with topic sentence.
32. Connect paragraphs with transitional phrases.
33. Provide or clarify principle of development, e.g., description, argument, cause-to-effect, example, details, expanded definition, space order, time order, explanation, familiar-to-unfamiliar, comparison or contrast.
34. Rearrange paragraphs or their parts.
35. Make paragraphs(s) shorter——longer.
36. Write a summarizing paragraph or so.
37. Rewrite along these lines.

IF WRITER AND EDITOR ARE SEPARATE

Whether the reader of this volume becomes a writer or an editor will depend on well-established procedures where he takes a job; chances are he will suddenly be one or the other, with little to say about it.

Whatever his work turns out to be, he should thoroughly understand how writing and editing are related. He will come to see that the work of an editor takes up where that of an author leaves off. In receiving the writer's rough manuscript, the editor first decides if it meets minimum standards of organization and legibility. If not acceptable, it is returned for rework and clarification.

If an editor, he may have delicate and difficult relationships with the writers whose work he edits. Here are a few rules that should smooth his way. He should not make changes in a report unless he can give sound reasons for so doing. He should see to it that the writer understands the reasons for such changes; he can in fact merely suggest such changes to the writer and let him do the necessary revising. He should make allowance for the author's own writing style. But if adverse criticism of such style is necessary, it is more tactful for an editor to discuss, if possible, another report already issued.

But what if one becomes a technical writer and has an editor? What should the writer do or not do to smooth his relationship with this editor? One thing has already been indicated: the writer, with the help of guide sheet and checklist, should talk over his report with his supervisor. This will result in a better report and, thus, less friction with an editor. Second, the writer should read his report over carefully before giving it to an editor and should not depend on the editor to correct grammar and spelling; it is assumed these will be correct. Third, the writer should not become angry about changes made by an editor, but should confer with him and try to see his viewpoint. He should also express appreciation for the times the editor rescues him from errors of various types.

There may, then, be three persons closely involved in writing a report: a person we have called the commissioning supervisor for whom the report is immediately written, the writer or writers, and an editor. The reader of this book may become one of the last two or he may become both; that is, he may be his own editor. He cannot know beforehand just what the writing relationship and duties will be, but he should be prepared to adapt himself to the situation.

Supervisor, writer, editor: this is the group that must often work together. Nothing is more destructive of pleasure in one's job than to have continuing friction in the group. This can happen in many ways. The supervisor may do more editing than he should, either because he has a "rewrite complex" or because there is no editor, or no good one. On the other hand, the writer may be convinced he performs perfectly and may refuse to accept good writing counsel from either supervisor or editor. Still again, the editor may be too nit-picking, or he may be so ignorant of the subject he edits that he changes basic meanings. Friction and unpleasantness may occur in these and other ways—all of which, however, are avoidable if thoughtfulness and fairness are applied to the writing task.

EDITORIAL STYLE

The new employee may be asked to make the reports of his company consistent, or he may find them surprisingly inconsistent and may want to correct the situation himself. This may necessitate making many changes, since inconsistent practices can total a dozen or more, not just one or two.

Probably most of the remarks on style that are familiar to the reader have referred to literary style and do not pertain to technical writing.

Literary style refers to the distinctive ways in which a good writer expresses himself and to his personal and perhaps subconscious blend of methods for achieving any number of values, from clarity to dramatic effect. Literary style includes the writer's characteristic uses of language, such as fondness for certain sentence rhythms and forms; individual ways of choosing and presenting subject matter; and unique use of phrases for their connotative and suggestive effects. Analysis of literary style, although it can yield a certain amount of knowledge, is limited, and ends in mystery. One cannot understand the basic nature of a great literary style any more than one can comprehend the uniqueness of a great personality.

On the other hand, editorial style—or lack of it—is readily apparent to readers and easy for them to analyze. It refers simply to set ways of dealing with certain details of expression such as spelling, abbreviation, headings, capitalization, punctuation, indentation, and so forth. Each of these topics is governed by rules, and style books explaining them are available in many newspaper offices and other publishing organizations. In technical writing and editing (as in newspaper work), the style is usually carefully prescribed in some detail, being dictated by company and other specifications or by standard operating practices. Each company's usage may reveal peculiarities of its own. As Kipling writes, ("In The Neolithic Age") "There are nine and sixty ways of constructing tribal lays, and every single one of them is right." But in any one tribe, its usages are law.

So, the technical writer should follow any specifications that pertain. Previous publications can also guide him. The more objective he is, the less "literary" and subjective, the better.

Personnel concerned directly with editorial style and consistency are not only writers and editors, but artists, typists, printers, proofreaders, and supervisors.

No kind of style can exist without consistency of expression. Securing this is no big problem for literary stylists, since a fine writer tends naturally to keep writing in a certain way. But in the case of editorial style, consistency must be sought deliberately.

One way to be consistent is to follow "usage." Toy battles rage as to whether usage—that is, the way people actually use language—is determined by critics and their rules or by "the people." The latter is the stronger view: critics merely note how people talk and write, but cannot control it.

The area covered by usage is vast, being that of the language itself. A few examples, out of hundreds that people argue over, will show what is meant. (1) Should one say "I agree to the plan" or "I agree with the plan"? Answer: "to." (2) Should one say "not as tall as" or "not so tall as"? Answer: "so." (3) Should one say "identical to" or "identical with"? Answer: "with." People argue heatedly about such matters, even about the exact pronunciation of words. There are books and informed persons who tell us what the (average) usage is.

Editorial style itself divides into two parts, one dealing with *format,* the other with *terminology.* A third subject to be treated because of its importance, *punctuation,* is also part of literary style. See "3. Punctuation."

1. *Format*

This concerns itself with such matters as margins, headings, indentation, pagination, and type sizes and faces. These topics may seem trifling to the nonprofessional and are indeed among the least interesting of our topics. But in each instance the point is that since the margins, and so forth, will appear in one way rather than another, this way might as well be correct and pleasing. Why, for example, have a 1 1/2-inch margin down the right-hand side of every page, if a 1-inch margin looks better? Or why have the pages inconsistent in regard to margins?

All of editorial style concerns itself with matters like this, which should be faced at the outset and firmly decided upon. It is the part of wisdom not to ignore these style problems, but to learn what they are and how to solve them. Otherwise they will probably remain a bore and a nuisance throughout one's writing days.

Margins. Provision must be made for consistent margins if the publication is to have a professional appearance. Sufficient space for page headings, page numbers, and for binding is, of course, essential. Remember, too, that the margin along the bottom of the page should be slightly wider than the one along the top, for good aesthetic balance.

For the 8 1/2″ × 11″ typewritten page, one of the most pleasing margin combinations is to leave 1-inch margins horizontally along the top and vertically down the right-hand side; and 1 1/2-inch margins horizontally along the bottom ("thumb-room") and vertically down the left-hand side.

Headings. The *order* of headings used for an outline is normally followed in a report or other paper that is written *from* or *to* that outline. That is, if I precedes A in the outline, it does so also in the written report, and so on for the rest of the ordering arrangement. The report conforms to the outline for it. This is easy to understand and seems only sensible. An inconsistency would be glaring: it would mean that, for instance, the topic that followed I in the outline would not be the same topic that followed I in the report.

We can easily spot such inconsistencies. But *all* inconsistencies of style should be avoided; even when they are not consciously noticed, they may well affect the subconscious mind of the reader, making the substance of the writing more difficult to grasp.

The problem of headings is of course by no means limited to writings that have outlines. Pick up a newspaper and note the many *varieties of headings* or headlines, and subheads, that are used. In several types of technical writing, headings are amply used. The problem of achieving variety and contrast with them then arises. This can be done in many ways if linotype machines and letterpress printing are used. If one is limited to the office typewriter, its capital letters may be contrasted with lower-case letters, and certain types of headings may be underlined.

But this does not dispose of the complexities to be considered in choosing headings. Their *position on the page* is also involved. There is, for example, the

Flush-Left Subhead

This may be followed, as a matter of style, by an indented sentence, such as this one. There is also the

Centered Subhead

This becomes a "boxed" heading by drawing a rectangle around it. Or one may use the type of heading appearing below and to the left, called the cut-in heading.

The Cut-In
Heading

A variety in headings can easily make written material more attractive and more easily read. If the technical writer is regularly preparing many columns of material, he should consider adopting, as part of his style, the practice of breaking up each column with one subhead for every three vertical inches of ordinary typed copy.

Indentation. This refers to the number of spaces from the left margin at which each paragraph begins. In technical writing that contains many subheadings, numerous tables, or mathematical and other formulas, particular attention should be paid to deciding upon and using a consistent style of indentation. Subheadings are ordinarily indented more than major headings, just as is done in preparing an outline.

Pagination. This means the system adopted for placing page numbers on pages. By referring to several books and magazines, note the variety of methods used. For a printed publication, the author prefers the style of numbering the left-hand page at the lower left corner, and the right-hand page at lower right.

Type face and size. These refer to the kind or appearance of the type used and to its height and width. The many choices possible are best displayed in so-called "type books." Some of these are dispensed free by printing establishments to those interested. The student should obtain one if possible, read the instructions and explanations it contains, and observe carefully the various type faces presented. He should note which type faces appeal to him aesthetically, and try to learn their names. The study of type faces can become a pleasing discipline, and can help to dispel some of the mysteries of printing.

The style of an organization can be expressed by the kind of type it regularly uses for headlines, such as Spartan, Alternate Gothic, Metro, and by the body type used, such as Caledonia, Garamond, or Bodoni.

The proper names of type faces just used refer to the *typeset* copy, which may be printed either by the letterpress or offset method of printing. When one is publishing a *typewritten* book and using offset printing, one has less choice. The electric typewriter (also the headlining machine) offers choices in kinds of type face supplied. Even the ordinary nonelectric office typewriter provides either "pica" or "elite" type. Not only are these two type faces different in appearance, but the pica measures 10 letters and spaces to the inch, horizontally, whereas the elite measures 12. The editor must remember, therefore, not to mismatch them (e.g., in revision work), but to use them consistently.

The height of type is measured in terms of "points" (72 to the inch), both width and height in "picas" (each pica is one-sixth of an inch in length). Two different kinds of type face, though the same height in point size, can differ in width and thereby look surprisingly different.

2. *Terminology*

Terminology concerns itself with such particulars of editorial style as *capitalization, abbreviation, compounding of words, spelling,* and *symbols.* Note that this division of style does not concern the rules of grammar, knowledge of which is presumed.

One of the commonest errors in technical writing is the inconsistent use of terminology. This is because of a failure either to decide upon a standard of style or to follow it. When, for the sake of clarity, the writer must repeat a term several times, he should be sure to use the *same* term. For example, if an "engine cradle" has been mentioned, all succeeding references should specify, if not "engine cradle," at least "cradle," and not "mount," "dolly," or "stand." The habit of being consistent is just as easy to form as that of being inconsistent—and is a necessity for effective technical communication.

Capitalization. Derivatives of proper names are usually capitalized, as "Elizabethan," unless they have acquired an independent common meaning, as "venturi tube." Common nouns or adjectives forming an essential part of a proper name are capitalized, as "Statue of Liberty." Trade names are capitalized, for example, "Kodachrome." Two common nouns or adjectives that form a proper name are capitalized, as "Western States"; if they are separated, they are not capitalized, as "western farming states." Webster's dictionary is, in general, a good guide for capitalization; but technical knowledge is advancing so rapidly that not all terms are in the dictionary. In such cases, discover what the preferred usage is in the particular field of scientific or technical knowledge and follow that usage.

Abbreviation. This practice saves space and patience, the extent to which it is employed depending upon the style of writing. A technical publication usually uses a great many abbreviations acceptable in its particular field; they must be carefully learned. Many technical writing specifications and technical societies follow the "Abbreviations for Scientific and Engineering Terms" of the American Standards Association. When an abbreviation that is not well known is first used, it follows the spelled-out word or words in parentheses, as "electromotive force (emf)," and appears thereafter in the abbreviated form only. Abbreviations of measurement should not be used if numbers such as "three" are spelled out. Thus, "three lb" would be wrong; "three pounds" is right (or 3 lb). Under present technical writing style, a period is not used after an abbreviation, as "lb," unless it spells a word, as "in." for "inch." Plurals of most technical abbreviations are written in the same way as are the singular forms, thus "13 lb."

Compounding of words. Compound words and expressions, those composed of two or more words joined together to convey single ideas, fall into three groups: those written as separate words, such as "book review"; those written with a hyphen, such as "brother-in-law"; and those written as one word, such as "motorman." Today's technical writing style is to minimize the use of the hyphen, writing in "solid" form all compounds using the prefixes *ante, bi, co, ex, in, inter, mid, mis, non, over, post, pre, pro, re, semi, sub, super, un,* and using such suffixes as *less* and *fold.* Some examples of these solids are the words antedate, bimonthly, comaker, deenergize, infrared, interface, midsemester, misstate, nonessential, postgraduate, reenter, semiautomatic, subcellar, transonic, unnecessary, prearrange, and postdate. Some of these solids repel one at first, for example "deenergize," but not after a little usage.

Spelling. English spelling is confusing because it is frequently illogical, and because so many words have two acceptable spellings; for example, either "gauge" or "gage" is correct. Since usage permits these variations, each writer and editor must make or learn certain decisions as to spelling style. As for general rules, with their many exceptions, a good source is the 17 rules in Webster's New International Dictionary. Fortunately, no good speller ever spells by rules; he spells by remembering how a word is written and how it looks when it is right. This skill can be greatly sharpened by noticing correct spelling and impressing the desired word images upon the mind.

The U.S. Government Printing Office, in its widely used "Style Manual," has compiled a list of words showing their preferred spellings.[1] Examples are adviser rather than advisor; aline, not align; ameba, not amoeba; ax, not axe; dialog, not dialogue; distributor, not distributer; flammable, not inflammable; liquefy, not liquify; plow, not plough; preventive, not preventative; programer, not programmer; stanch, not staunch.

Good technical writers are good spellers. This is because the unremitting care that they give to writing problems and niceties naturally includes the correct spelling of words.

Use of symbols (primarily letters and numbers). This is a highly detailed field, for which many aids exist, such as the manuals published by the just-mentioned American Standards Association.

Many mathematical symbols are letters, such as those of the original Greek alphabet. Most chemical symbols are composed of letters, such as Fe for iron and Al for aluminum. For these the approved list is that of the International Union of Chemistry. Letter symbols, including subscripts and superscripts (symbols below and above the typed line) should, ideally, be supplied by a typewriter, rather than being handwritten.

A few general rules on the use of numbers as such may be given. Current technical writing style ordains that if a number is less than 10 it is spelled out, except in a series including the number 10 or one higher, in which case all figures are used, and except for "unit modifiers" such as "4 sec," "5 lb," and so on. In these cases the unit is the "sec" or "lb," and the modifier is the "4" or "5." When the entire unit modifier is used as an adjective, as "a 4-sec pause" or "5-lb weight," a hyphen, as shown, should appear between the original unit and its modifier. When this entire unit modifier becomes still more complex, being then called a "compound modifier"—for example "*two* 4-sec pauses," or "*three* 5-lb weights"—the leading numeral is spelled out, as the italics show. Ordinal numbers are spelled out, as "first" and thirteenth." Numbers at beginnings of sentences should be spelled out, as should fractions occurring alone, as "one-half." Large numbers are best written with numerals followed by the appropriate words, such as "10 million" and "20 billion," unless they begin sentences.

3. *Punctuation*

To master punctuation is said to give an aspiring writer considerable confidence. If he feels he can punctuate correctly, he tends to feel he can

[1] The first page of this manual is reproduced in the Appendix, p. 169.

write. This is partly because, while the hundreds of thousands of words that exist all require punctuation, only about 10 "points" (punctuation marks) are used for this purpose.

A point has but one reason for existing: to make a sentence easier to understand. This must mean that it is itself a meaning. Points, then, are simply a small set of meanings added to words. What are these meaning*s*, and how did they develop?

Early writing had no punctuation! Everything was written "solid" and lookedevenworsethanthisdoes; but since the fifteenth century, both scribes and printers have used points. First, they saw the value of leaving spaces between words; it made for faster reading. Second, they marked off sentence units with periods, to signify the pauses used in speech. Third, they began slowly to use commas, semicolons, colons, dashes, and other marks which made writing more comprehensible. Although each printer had his own ideas of punctuation, standard practices gradually developed—since there were fewer printers than authors!

The rationale of (written) punctuation is that it too, like our written language, is based on language as spoken. To what do points correspond in the spoken language? Pauses, mostly; also pitches. A period indicates a longer pause than does a semicolon, which signals more pause than does a comma. A question mark indicates an ending voice rise. The phonetic linguists have interestingly worked out this correspondence between (part of) punctuation and (part of) intonation. (The latter subject comprises juncture, pitch, and stress.) But we must leave this insight to merely state some basic punctuation practices.

Comma. Instead of mastering some 17 explicit rules for use of the comma, it is more helpful to know that this punctuation mark is used in three main ways: (1) singly, (2) in pairs, and (3) with coordinating conjunctions (i.e., with "and," "but," "for," "or," and "nor").

(1) The comma is often used singly to separate two adjectives of the same type, or coordinates,[2] as in "a clear, well-written article," or "a brilliant, conscientious supervisor"; to set off certain adverbial clauses or introductory words and phrases, as in "If any acid can dissolve it, this should," or "Yes, follow the road"; to indicate an omitted word or words, as in "We acted graciously; our opponents, abominably"; to separate city and state, as "Albany, New York," or day and year, as "March 23, 1967."

(2) Used in a pair, commas (note underlines following) provide pauses, enclosing expressions that could be left out without destroying the sentence, as in "This machine, which has found universal use, can be bought for

[2] Oddly enough, if the adjectives are of different types, or are uncoordinates, as in "a clear official statement," they should *not* be separated by a comma.

$25." This use of the comma gives what are called nonrestrictive (nonessential) meanings to clauses in sentences, as "The colonel, who was at a nearby post, arrived promptly." Here, the colonel's whereabouts are indicated by the commas as *not being essential* to the meaning of the sentence. But if the commas are left out, the restrictive (essential) meaning results. "The colonel who was at a nearby post arrived promptly" indicates that only *that* colonel is intended as the subject of the sentence.

(3) Used with a coordinating conjunction, the comma indicates the approximate midpoint of a compound sentence, as "Shut off the valve, and then you can observe the increase in pressure."

Semicolon. The semicolon and the colon are the two remaining "interior" (used "inside" the sentence) punctuating marks. The semicolon, which is less emphatic than a period but more so than a comma, is often used to separate phrases or clauses that themselves contain commas or that should be set apart more sharply than usual, as "George Washington, our first president, was a soldier; and so were other chief executives, such as Grant, Jackson, and Eisenhower." The semicolon is the proper punctuation to use between independent clauses in a sentence when these are not connected by a coordinating conjunction. Thus, "He has done brillant work and so his promotion is probable." But without the "and," one would write, "He has done brillant work; his promotion is probable."

Semicolons should be placed before "sentence connectors" in sentences. The sentence connectors (linking independent clauses that could *serve* as sentences) are such words as the following: however, consequently, nevertheless, therefore, accordingly, indeed, in fact, thus, hence, moreover, furthermore. For example: "He was told to make inquiry at once; therefore his employment could begin today.

Semicolons should also be placed before "specifying phrases" such as "that is" and "for example." So we would write, "He was put in his place; that is, officially reprimanded."

Colon. The colon is used (note underline again) to introduce a fuller explanation of what immediately precedes it: as "We require four items for this experiment: water in a clean dish, some talcum powder, a stopwatch, and a wet bar of soap." The colon usually heralds or precedes a listing, but always functions as a request to the reader to note what follows.

Dash. The dash often follows a listing and is followed by a summing-up phrase; thus it works in exactly the opposite way from a colon. For example: "Nuts, bolts, and brackets—these were the items he bought." The dash also marks the breaking-off of a statement (in literary dialogue mostly), or a stressed phrase that follows it and marks a sudden turn of thought. "He counted slowly and deliberately to ten—then he let me have it." (On the typewriter, the dash is made by striking the hyphen bar twice.)

Hyphen. The hyphen (mentioned briefly under compounding of words) is a combining mark, with several important uses. Interestingly, it joins words *before* usage brings them solidly together: basket-ball, before it became basketball; to-day, before it was today; tax-power, man-power, and so on. It separates compound numbers (twenty-one years), fractions written out as adjectives (two-thirds majority), and compound fractions (twenty-one twenty-fifths of those voting). The hyphen separates two or more compounds with a common base (1-, 2-, and 3-inch nails). The common base referred to is "-inch nails." It is also used to break words at the ends of lines. The basic rule for dividing words is to do so between syllables. This point of division is usually after vowels, as "di-vi-ding."

A hyphen should be used if meaning might otherwise be misunderstood; for example, write "re-form" (to form again) to differentiate it from "reform." Also, many compound words are still written with a hyphen, and these must be learned individually rather than by rule. Examples are anti-Semitism, by-product, half-truth, and vice-president. One hyphenates two or more words used adjectivally for description preceding nouns, such as "long-awaited reform."

Apostrophe. The apostrophe is used to indicate where letters have been (quite correctly and grammatically) omitted in contractions, such as "doesn't." They are also used to form, alone or with the letter *s*, the possessive of nouns, as "ladies' " or "man's", and to make plurals from certain terms including artificial compounds, as "all the f's were made backward," or "p's and q's." The latter are called "coined plurals."

Two miscellaneous rules of punctuation may be noted. (1) When a request is made as a formal question, a question mark is not needed, as "May I request a prompt answer." (2) At the end of quoted matter, periods and commas are placed *before* quotation marks, whereas semicolons and colons are placed *after* quotations. Examples are:

> The title of this essay, "Plants Hold The Basic Patents," arouses one's interest.
>
> Patrick Henry said, "Give me liberty or give me death"; a most courageous utterance.

Punctuation is such an important subject that it deserves special treatment. In the following test on punctuation, please answer each question either from memory or by using the text. Then check with the answers in the Appendix, p. 208.

Exercise 23. *Punctuation*

1. When does one place a period after an abbreviation?
2. Why does the mastering of punctuation tend to give confidence to a writer?
3. What are points, or punctuation marks?
4. Should one use commas between coordinate adjectives? Between uncoordinate adjectives?
5. Can the use of a comma ever indicate the omission of words?
6. How can one show that certain phrases and clauses in sentences are not essential?
7. How can one separate independent clauses without using the coordinating conjunctions such as "and" or "but"?
8. How are semicolons used with sentence connectors?
9. What point does one use to *precede* a specifying phrase such as "for example"?
10. How is the colon a "mark of anticipation"?
11. To stress a phrase, what point or mark would one use before it?
12. What is the basic rule for dividing words?
13. Should one ever divide a word after a vowel?
14. What indicates a grammatical contraction? A coined plural?
15. How are colons and semicolons used in relation to closing quotation marks? Is it correct to put either a period or comma inside a closing quotation mark?

ARE YOU ALSO THE PUBLISHER?

The writer or the editor of a report may find, probably to his dismay, that he is also the publisher and must coordinate the publication processes. If so, he should understand these processes well enough (1) to know who is responsible for what, and (2) to talk their language, thus facilitating the work.

The new employee may be surprised to learn just what his duties are. He may be summarily asked to fill an existing need. For instance, he may have to order, edit, and be responsible for the necessary graphic aids from the art department. Certainly somebody has to do this. Since a report is intended to be circulated, copies of it must be prepared by a suitable method; this, too, somebody must do. The reader of this chapter, perhaps now only a student, may suddenly find himself superintending the preparation of typed material for reproduction.

1. *The Flow Chart*

It may, therefore, be helpful here to study a "flow chart" of a typical publishing procedure. Note that the technical editor is "in the middle" most of the time, with his hands and head full of "copy" (text or art for publication) and technical considerations.

Large businesses that prepare and publish their own reports, employing their own editors, usually use the photo-offset method of printing. This allows revisions to be made more easily than does letterpress, is generally cheaper, does not require type to be set but uses an office typewriter, and handles illustrations (art) economically.

The photo-offset method requires that perfectly typewritten pages be photographed, with printing plates then being made from the resultant negatives or directly from the pages. The typed pages are usually prepared within the business concern itself (i.e., "in-house"), and their readiness becomes an editor's responsibility. He may therefore deal professionally with typists as well as with artists; whatever art is needed is also photographed by the offset method.

Before commenting on Steps 1 through 9 of the following flow chart, let us note particularly the arrow between 1 and 2. This phase of the complicated publishing process of getting out a report has been discussed in connection with the "Guide Sheet for Writing Assignment" (p. 108), the "Supervisor's Checklist" (p. 122), and in "If Writer and Editor are Separate" (p. 124).

Let us state again that the technical writer prepares the text copy—typewritten, let us hope—and also prepares, in rough form at least, whatever "art" (sketches, photographs, etc.) is needed. If he and the editor are not one and the same individual—that is, if the original writer does *not* also do the editing—the writer should realize and abide by the following truths and practices: (a) The editor is not supposed to have the technical knowledge of the writer, from whom accuracy, conformance to fact, is required. (b) If text changes are not understood, the writer should calmly discuss them with the editor and listen to his reasons for the changes. Both writer and editor can frequently profit from such conferences. (c) Where the editor has changed· the meanings of words, it may be because the writer, by unskillful usage, said what he did not mean. In each case, it is the editor's responsibility to confer with the writer. The editor may be quite right to have changed the meaning, but he must be sure about this. (d) Use by the editor of either *more or less words* than the writer used may have increased clarity or some other desired quality. (e) If the writer finds his original draft or manuscript hard to understand when re-

Flow Chart of a Typical Publishing Procedure

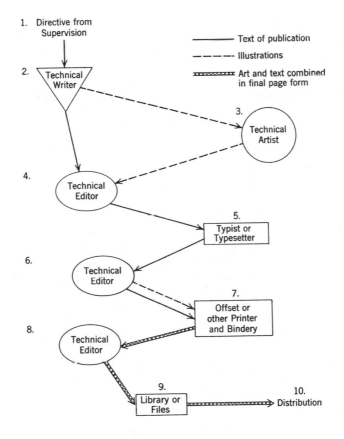

turned for his approval, this may be merely because he does not comprehend the editor's symbols, the marks used in preparing copy for the typist or printer. (More on this later.)

Let us now turn specifically to Steps 2–3, the relationship between the technical writer and the technical artist. The preparation of art for a publication is a complicated subject in itself, partly because types of art are so various. For example, if the writer wants to use photographs, he may be able to obtain copies by giving the correct "neg numbers" to the reproduction department. But for a line drawing which does not yet exist, he may have to prepare a sketch of it, complete enough for an artist to work from. The writer usually gives all the rough art, in whatever form, to the editor, who checks it over and sees that there are enough of the "makings" for a professional artist to work with, then confers with the artist. Although the latter may do all of the art work, the editor usually has final responsibility for it.

The most important thing to know when dealing with technical artists is not what they know, but what one's relationship with them is—or who is helping whom. *Sometimes* the editor in control is himself an *art* executive. In this case, the writers and editors are making their contributions to the art department, which has the responsibility of producing satisfactory end-products (such as advertising brochures). But in most cases, at least for manuals and reports, editors have the final responsibility, and therefore the art department is working to *their* requirements.

Either situation means that the editor must be able to talk the language of the artist. This is no mean feat. Whoever acts as liaison with the artist should certainly earn his respect and cooperation by giving him the proper information or instructions and—something ignored in treatises on technical writing—must usually *edit what the artist produces*, making sure that it communicates efficiently. This responsibility for the art work is always exacting and can be troublesome.

The relative importance of illustration and text in a report or manual depends on the individual case. Some units of technical writing have no illustrations, though this is rare. Others could not be understood without illustrations. Examples might be photomicrographs of crystal structure, photographs of broken specimens in tensile strength tests, and schematic drawings showing the interconnections of parts in electronic devices. Speaking generally, illustrations save words and supplement them, but almost never supplant them. Art should therefore be used to *assist* the text.

Completed art work should always be thoroughly edited. The need for this is seen by reflecting that art and written copy are usually handled separately. When they are united in finished form, they should be consistent—something that only editing can ensure. For instance, *any* words that are on the art itself should agree with what they refer to in the text. As a glaring example, such a "callout" as "wheel" on the drawing of an airplane should not be termed "landing gear" in the adjacent text. Artists are not trained to catch such discrepancies; editors are and must amend not only all such discrepancies in nomenclature, but also mistakes in spelling.

The responsibility this puts on editors is best met by making sure that artists receive properly edited material to begin with. The lack of this is what artists rightfully deplore. To remove the fault, writers, editors, and artists should confer at the very beginning of the art work, to make sure that copy presented to artists for drawing—for finalization, that is—can be understood down to the very last line and letter. What Greek letter is this on the drawing? Is that a subscript or a superscript? Is this term to be capitalized or not? Can we reduce this drawing in size? Is it too detailed, too "busy"? To thrash out, as a first step, all such questions means to save much sad discovery later that things have been done "wrong," necessitating costly, time-consuming corrections.

If the editor does not get, in his rough manuscript, a list of illustrations showing how much art, how many figures there are supposed to be—and he usually does not—he must prepare such a table. He must ascertain from the author how much art there is supposed to be, then check to see whether he has it all; if not, he must make plans to obtain what he lacks.

When the artists have drawn all the figures that are to be used, they must be shown by the editor to the author for his final approval—a step usually taken when the perfectly typed text is also ready to show the author. He remembers that both text and art are to be photographed, and that they sometimes appear on the same page. When this finished art is already the right size for the page, it is often pasted right on the page, in the exact space planned for it. This entire page or "repro" (page to be reproduced) is shown to the author for his approval. But sometimes the art must be reduced in size (by the camera, in being photographed) by a certain amount predetermined usually by the editor, before it will fit into the space planned for it on the page. In such cases, the reduction is usually not done until the entire job is sent to the printer. But the author realizes it *will* be done. He sees the art to be reduced, he sees the space left for it to be "stripped in" (the negative or the reduced art is pasted into the negative of the rest of the page), and he sees whatever text may have been typed on the page.

From these visible signs and facts, the writer or author evaluates the entire job and—perhaps after asking the editor to make some last-minute changes—he "signs off" the job, his signature on a work order meaning "Okay; go ahead and print it." The signer then does not usually see the job again until it comes back to him as a printed report, brochure, manual, or whatever.

Step 3, then, includes the return of the finished art to the editor, *who holds it* until the finished typing is also ready to show to the author. Before this formal presentation is made, Steps 3–4 and 4–5 must be completed. The editor, *after editing it* (see next paragraph), must send all of the text that is to be typed to a typist. And she must return it (Steps 5–6).

The editor should plan so that the work in the art department is proceeding concurrently with his own editing work, which is as follows: He checks to see that the text copy or rough manuscript (often called "rough draft") is complete (just as he did with the art). He then edits it for conformity with any specifications that pertain and for correct grammar, spelling, and consistency of style; he may confer with the writer at least once to get his okay on important changes.

Part of what the editor does will of course depend on the type of publication being planned. An engineering report may require a table of contents and an abstract, whereas a brochure will have neither. A manual would have a table of contents, but no abstract.

As the editor reads and marks up the original manuscript, he can also use the margins on each side of the page as follows: (a) When figures (i.e., pieces of art such as drawings, photographs, charts, graphs) are first mentioned in the text, it is usually advisable to have them printed immediately following this mention so the reader can get maximum value from them. The editor can alert the typist to their planned appearance by writing in the left-hand column of the text the final dimensions of each piece of art so she can leave enough blank space on the repro page for it. (b) The editor can use the right-hand margin to write queries about the text that he must discuss with the writer. This discussion should probably take place just before the editor gives the manuscript to the typist.

The typist (or possibly typesetter) who receives the edited manuscript from the editor may be either inside or outside of the company. At this point, we face a varying situation that can only be sketched here. Some companies, especially the largest, have not only their own repro typists, but also their own complete printing units. Heavy work loads, however, may at times force even these companies to send their printing work out to "vendors." These situations, varying with work load, plus the fact that the editor may regularly prepare various *kinds* of publications, put a great premium on his professional knowledge and his ability to deal smoothly with different associated groups. He is supposed to understand all the jargon of these groups. Otherwise, he cannot do his job properly.

The reader may still be wondering why an editor would ever give a rough manuscript to a typist, since it is presumably already typed. This is because he finds corrections to be made, and because repro typing is itself a means of printing. Electric typewriters with "executive" or "book-face" type, for example, are often used to prepare perfect copy from rough manuscript given to typists by editors.

By "perfect" is meant "ready for photographing." As a step in the photo-offset printing process, photographs of these repro pages are made, permitting metal or lighter-weight plates then to be made from the resulting negatives, or made directly from the repro pages. Finally the printing is done from the smooth-surface plates.

Comparison of the offset with the other leading printing method, letter-press, would show that, although basic editorial processes remain the same, the routing steps and liaison knowledge required are unique for each.

Remember the usual situation in large plants: use of an electric typewriter for typing, and offset method for printing. So, the layout typist (Steps 5–6) returns to the editor (*with* the rough manuscript or draft), the new freshly typed repro pages ready for photographing. But *are* they ready? It is one of the editor's duties to check every repro page and make sure it has been typed correctly from the rough manuscript. (If he finds mistakes, they may be so slight as to permit the typist to white out and type over the same place; or perhaps a new page or pages must be typed.) In any case, the editor is responsible for final correctness.

2. *Proofreader's Marks*

This process of comparing the fresh repro pages with the rough manuscript is called proofreading. The operation should not be slighted; accurate copy is a "must" in technical writing. The comparing is usually done with two persons; one may read aloud the newly typed text as the other person reads it silently from the rough manuscript. The two texts should agree. If they do not, and the new, typewritten text is at fault, it must be amended.

The editor's written mode of indicating mistakes will depend on using either proofreader's marks or some kind of correction sheet. Both methods are shown here.

The mistakes that the editor may find on the repro page are of two general types: *errors in style* and *textual errors*. Failure to capitalize a certain type of subtitle, or to add page numbers, would be examples of the first kind. *Omission* of a paragraph from the rough manuscript, or *misspelling*, would be examples of the second kind.

Both *copy preparation* (marking up the text for typing) and *proofreading* (comparing what is final-typed with the original manuscript) require that certain marks be used as a means of communication among editors, typists, and printers. The more than 50 of these proofreader's marks are not all presented here because they are used not only limitedly but with many variations from office to office; also, for work with repro pages, a neater method exists. Only 15 follow, then the "neater method" is explained.

These marks, indicating desired text changes, properly appear in pairs: once *in* the line of copy itself, and once in the margin of that line, to explain. In some cases, the marks of a pair are the same. To avoid this repetition, one can merely make an arrow in the margin.

If or when an editor finds that he needs to use proofreader's marks in his work, he should compile a *suitable set* and have it clearly approved by, and distributed to, all concerned. (Complete sets of such marks often appear in printer's type books, dictionaries, and in various works on publishing and printing.)

Marginal sign	*Mark in text*	*Meaning*	*Corrected text*
ℐ (or ℓ)	castings	delete	casting
⌒	die maker	close up space	diemaker
#	online at	separate or leave space	on line at
¶	follows. Next	start new paragraph	follows. Next
no ¶	done carefully. Plans have been	no new paragraph	done carefully. Plans have been .....
⌐	0.0134	move to right	0.0134
⌐	A. INITIAL DATA . .	move to left	A. INITIAL DATA
tr.	extended was	transpose	was extended
⊔	Corrections in	lower	Corrections in
⊓	text are often	raise	text are often
Sp.	The (3rd) trial	spell out	The third trial
u.c.	The supervisor	capitalize ("upper case")	The Supervisor
l.c.	This Trial Run	uncapitalize ("lower case")	This trial run
stet	We think that the	retain	We think that the
⊙	end But final	Insert period	end. But final

In case of photo-offset printing, the editor may state what changes he wants made, using the proofreader's marks, *on the repro pages to be photo-graphed*; he uses a pencil that makes light-blue marks because these marks will not photograph; they will merely tell the typist what changes she is now to make on the repro pages so they will be perfect for photographing.

3. *The Correction Sheet*

The neater method avoids making marks on the repro pages even though the marks will not photograph. The repro pages to be typed on are *pre-printed* in that same blue that "drops out" when photographed. These pre-printed pages are numbered (in faint blue) vertically down the left-hand side to indicate lines; they start at the top and are about 1/4 inch apart. They look something like this:

0- - - - -
1- - - - -
2- - - - -
3- - - - -
4- - - - -
5- - - - -, and so forth, down to about Line 37. The repro typists do their typing on these otherwise blank pages.

Besides this preprinted repro page, there is a second form to be used, a *correction sheet*. These sheets are printed up in ordinary ink. On them the editor indicates, by naming page number and line number, every mistake he wants corrected, as well as the correct version.

Use of the correction sheet avoids making marks *on* the repro pages. The sheet, with a few examples of changes desired, looks like this.

CORRECTION SHEET

To Author _____   Date _____
From Editor ____   Report No. _____

Page	Line	Was	Should Be
29	2	effects	affects
35	4	feul	fuel
36	21	g/cm^3	gm/cm^3
39	8	stream	steam

The editor, in proofreading the typed repro pages against the rough draft (the original manuscript), spots any mistakes made by the typist, then indicates on the correction sheet, by page and line, the changes he desires. He gives this sheet, together with the typed repro pages, to the author, so that he can see what changes the editor means to make and not duplicate them. For the author proofreads the repro pages too, and using the same or another correction sheet, adds his own desired changes. After the typist has made the editor's *and* the author's desired changes, the repro pages are presumably perfect and ready to be photographed.

This is no time to rewrite the manuscript. Any but minor changes may foul up the page layout, require pages of retyping and new layout pages, repagination, changes in the table of contents, and result in missed deadlines and unpopular writers.

The art, too must all be ready for photography and must go along with the rest of the copy. For art, too, is copy (i.e., material to be copied). Both editor and author must see the work of the artist and agree that it is correctly finalized from the "rough art" that the author originally gave to the editor.

4. *The Printshop Checklist*

At this point, at the end of Steps 5–6, we suggest that the editor use a "printshop checklist," which he will probably have to devise and have printed up for himself. This is a form that lists every feature the report should possess. This list of features will vary from one company to the next, so any list we indicate here merely outlines the idea. See next page.

Mention of all the things to check on and verify before giving the report to the printer (Steps 6–7) may take up more than a page. Less than a page has been used in this example, but no spaces were left for checking correctness of distribution, references, and other matters which one may wish to add.

Such a checklist will have about as many versions as there are editing centers. But working up a good one and using it will save many mistakes, many embarrassments, and much expense when one has to decide whether to reprint an entire issue because of an editor's failure to correct some error before giving the job to the printer.

Steps 7–8 have variations too. What is printed is not always, perhaps not usually, returned to the editor; it may be sent directly to the company library or other agency entrusted with the mailing out of a certain number of the reports. The editor may or may not be responsible for Steps 8–9–10, which are shown on the flow chart for the sake of completeness. Clearly the report has not served its function until it has been distributed according to a "clean" (free of wrong names and addresses) and up-to-the-minute mailing list.

Printshop Checklist

Cover Page
  1. Official number
  2. Title
  3. Security classification
Back of Cover Page
  1. Legal notice
  2. Price
General
  1. All text pages have page number
  2. All text pages have report number typed in at bottom center
Title Page
  1. Official number, same as on cover
  2. Total number of pages shown
  3. Title same as on cover
  4. Author(s) name(s) appear correctly
  5. Contract number is correct
  6. Issue date is correct
Table of Contents
  1. Every section listed exactly as in text
  2. Captions listed exactly as on figures and tables in text
  3. Figure numbers assigned in order of appearance in report
Illustrations
  1. Numbered consecutively
  2. Each has a caption

Not only are there many different publishing operations in use, but any particular one may change radically with time and new methods. So an editor should be adaptable, versatile, able to acquire new skills. Since he makes many highly specified, detailed moves, he must be accurate and must either have a good memory or be good at making and keeping records. He should be able to work with efficiency under the pressure of meeting deadlines. He should get along with all kinds of people. He is frequently held responsible for publications appearing in correct form and on time.

We have briefly considered several situations, knowledge of which is helpful in turning out reports. Since today's student does not know at exactly what point he may be involved in the publishing process, we have traced it from beginning to end, from "report situations" to final distribution. Thus the student will have some perspective about the entire publishing procedure, as well as about his particular niche.

SUGGESTED FINAL FOR CHAPTER IV

The student will benefit from writing out answers to these 20 questions. He may then turn to the Appendix, p. 209, for suggestions.

1. Considered in the simplest possible way, what is a report?

2. How can one know when he is ready to begin writing his report?

3. What poor situation does use of the guide sheet overcome?

4. Name the possible elements in a report.

5. What is usually the most difficult part of a report to write? Why?

6. Name four or five first-order headings that one might find in the *body* of a report.

7. How does the report of a scientific object differ from that of a scientific process?

8. How can one avoid committing the most common faults in reports?

9. What considerations determine whether to use the active or the passive voice in a report?

10. After a report is written, what major benefit is derived from using a checklist?

11. State two rules or practices that should help an editor work effectively with a writer. State, conversely, two rules or practices that should help a writer with an editor.

12. Explain briefly the three topics discussed under editorial style.

13. Give the main rule covering the making of compound words with the regularly used prefixes.

14. What general advantages does the photo-offset method of printing provide?

15. Reproduce from memory the flow chart of a typical photo-offset publishing procedure.

16. How may an editor use to advantage the left-hand and the right-hand margins of the manuscript he is editing?

17. What is proofreading? How is it done and why is it important?

18. What method was recommended for having mistakes in typing corrected?

19. What is a printshop checklist, and what is the advantage of using it?

20. What qualities should a good editor possess?

V

Writing to Spec

It is important to see why an expanding civilization naturally generates specs. These function generally as directives for new technical knowledge. Specifically, they help us to master many processes of manufacture, from "hardware" to written products themselves. The use of specs is made easier by studying the major types and some of their interrelations. To write a spec itself is perhaps the ultimate in responsibility and care for a technical writer. There are several helpful ways to go about this task, all of which can serve.

AN EXPANDED DEFINITION

A specification is a clear and accurate description of the technical requirements for (1) handling a material or (2) providing a product and (3) a process or service. The spec includes the procedure by which it can be determined that the requirements have been met.

Thus a specification, henceforth called a "spec," is an instruction or set of them.[1] Civilization is filled with specs, since they represent the wisdom of the race. Parents communicate unpublished specs continually to their children. The "Ten Commandments" is perhaps the most famous of all specs for living life successfully and ethically.

These pieces of advice are found at all levels of endeavor. Take, for example, the following excerpt from a section on "Gears" in an Engineering Design Manual. (Many large manufacturers have such multivolumed depositories of how best to do things.)

"It is recommended that special gears (including worms and racks) that are to mesh together be shown on the same detail drawing, to preclude malfunctioning as a result of differences in methods of producing gear teeth. Noninterchangeable gearing may thus be given the notation 'KEEP IN MATCHED SETS.' "

[1] The first page of MIL-P-38790 is reproduced in the Appendix, p. 171.

147

This fruit of bitter experience, that drawings of meshing gears should travel in pairs, is an insight worth preserving. So are all competent specs, examples of which are the U.S. Constitution, the California Motor Vehicle Code, city building codes, postal regulations, college entrance requirements, parliamentary rules of order, contracts, insurance policies, all rules for sports, recipes, the National Electrical Code, JAN Wiring Color Code, American Standard Code of Pipe Threads and Fittings, Society of Automotive Engineers Standards, and Rules of Practice of the U.S. Patent Office.

Part of anyone's education consists, then, in learning *some* specs. After realizing their ubiquity and usefulness in life, we are the better inclined to consider their function in technical writing.

THE IMMEDIATE AND GROWING NEED FOR SPECS

Things that are (1) built must be (2) assembled, (3) installed, (4) inspected, (5) operated, (6) maintained, (7) repaired, and (8) overhauled, with the help of (9) parts lists. A type of spec corresponds to each one of these activities. Considering that not only hundreds of private industrial giants but also the federal government, with its defense and other activities, are variously engaged in the above operations, we can see why thousands of specs exist and must be written. Many of them must also be periodically revised.

Let us see clearly how specs naturally mushroom in a new field. Take the Space Age. For space apparel, there must be requirements of thermal insulation, ventilation, vapor-diffusion resistance, and air permeability. These conditions will be expressed in rigid specs that somebody must write. For space cabins, spec requirements will cover heating, ventilating, and air conditioning. The double-hull structures needed for some space vehicles will require new strength-to-weight construction standards, carefully specified by such authorities as the American Institute of Chemical Engineers. Materials technology will find new standards (specs are standards) involving rocket fuel analysis, combustion efficiency, and the measuring of physical and mechanical properties. The new refractory materials needed will have specs covering chemical resistance, electrical insulation, and mechanical strength. Specs will be needed to cover the new refractory materials: borides, nitrides, carbides, silicides. To achieve desired properties of metals we need to follow specific test procedures—to be found in specs—for applying tremendous heats within seconds. Guided missile technology will require greater standardization of missile ground support and of precision measurements covering tolerances for bearings, gyroscopes, fuel injectors, and transistors. A space vehicle may consist of a million parts made by hundreds of different manufacturers; yet all the components must be "up to snuff" and must work together in an integrated system of unfailing reliability. Specs must express this. The problem of reentry alone is creating specs that exist now only in the minds of engineers. Someday all the needed specs will have official numbers and will be printed on paper. Consider just those mechanisms called "compacts." For earthmen to conquer space, they must receive steady streams of electrical impulses translatable into precise knowledge about far-distant conditions. Compacts, of which there may be hundreds or even thousands in space regularly, are designed to provide steady electrical power for sending satellite and free-travel messages to Earth, and bouncing the messages about the world in our new global communication systems. These compacts must be made so they will start up only when in orbit, work gravity-free and without maintenance for at least a year regardless of sun, shade, spatial orientation, micrometeorites, and tumbling, and then dispose of their own radiation dangers upon reentry into the Earth's atmosphere.

Sooner or later, specs on all this so-called hardware, from airborne lifeboats to artificial satellites, must be written. Otherwise workmen would not be able to build them all unerringly, and in great numbers.

That chunky Space-Age paragraph just preceding should help to make us realize that we will always have specs. We can see why. Standardization in the space and atomic ages—standardization in any age—requires and *means* increased uniformity, performance, simplification, reliability, interchangeablity. These are ultimately expressed in black marks on paper— words, written by technical writers, and called specs.

One need not be a technical writer to be in a position to appreciate specs. Products that we buy as ordinary consumers and that must be assembled sometimes lack adequate directions. This simply means that the little *assembly* spec that was included in the purchase has failed to do its work. The children's game you just bought for a birthday present lacked clear step-by-step directions or failed to identify the playing pieces. An electric barbecue gave you directions for assembling the oven unit, which you did; then you found that connecting it to the stand required you to disassemble it. Or you bought a Geiger counter accompanied, you saw with pleasure, by a good circuit diagram and parts list. But you later found that the operating voltage was not stated, and this omission cost you many hours of research. Yes, specs are everywhere; and they should all be accurate and should all communicate.

HARDWARE SPECS AND WRITING SPECS

The subject of specs is, however, quite complicated. There are many thousands of them, of several different basic types. We can best thread our way through this jungle by keeping the point of view of the technical writer. How does this jungle of specs appear to him?

It seems clearly divided into specs dealing with (1) hardware or with (2) writing.

1. Hardware specs give directions for the making and upkeep of manufactured articles or for providing services. The contents of hardware specs are classified into commodity and process.

Commodity specs themselves divide into three kinds, of increasing structural complexity: materials, products, and equipment. Material specs give complete requirements for buying or handling the basic raw or semifabricated materials used in general construction and manufacture. Product specs give requirements for parts, subassemblies, and units used in equipment components. Equipment specs are concerned with complete units such as vehicles, standard kits, earth-moving machines, power transmission systems, and so forth.

Process specs give full requirements for providing services. Some of the widely followed procedures in process specs are better known and properly issued as standards.

All these commodity and process specs are cross-divided into "general" and "detail" specs. General specs are written to avoid repetition in the detail specs (more later on these).

Finally, a hardware spec can concern either performance or design.

Performance specs explain the required output, function, or operation of items or assemblies, leaving to the producer's option the matters of design, fabrication, and internal workings, for which specifications are not essential. The performance spec should state required rather than optimum performance.

Design specs contain the data needed to produce the item and thus include stipulations as to material, composition, weight, size, dimensions, physical and chemical requirements, and so on. Drawings are often used to establish exact features of design.

Hardware specs may stipulate that various chemical and physical properties be secured. Chemically, this may involve stating certain degrees of acidity or alkalinity, concentration, and composition. Physical properties will be interpreted in terms of tensile strength, hardness, specific gravity, shape, or other qualities in point. Measurements will be spelled out in such relevant terms as volumes, temperatures, tolerances, diameters, thicknesses, and gauge numbers.

An example of a hardware spec is MIL (for military)-H-5606A on "Hydraulic Fluid, Petroleum Base, Aircraft and Ordnance," approved by the Department of Defense for the Army, Navy, and Air Force. You see that we are terming hydraulic fluid "hardware"; this is justified because such fluid, in its system of tubes and valves, controls massive objects like airplane flaps, landing and retracting gear, and so forth. Paragraph 3.1 of MIL-H-5606A says: "*Qualification*. The oil furnished under this specification shall be a product which has been tested and has passed the qualification tests herein."

Paragraph 4.4 of the same spec, entitled "Test Methods," specifies that the pour point, flash point, specific gravity, color, and viscosity of the hydraulic fluid are to be tested by Methods No. 201, 1103, 401, 102, and 305.

To see how detailed these tests are, we read Paragraph 4.4.4 on "Evaporation": "A microscope slide shall be immersed in the hydraulic oil at room temperature. It shall then be removed and suspended by one end in an air oven at 65.5°C (150.0°F) for 4 hours. After removal of the slide and cooling to room temperature, the residual film shall be oily, and neither hard nor tacky."

2. Writing (or writer's) specs tell how to write such things as manuals and specs themselves. An example is MIL-M-38784A on "Manuals, Technical: General Style and Format Requirements."

Paragraph 3.1.6.1 says: "Manuscripts. The manuscript shall be complete in all respects, and shall contain all front matter, text, illustrations, and tables to be included in the manual as specified in the technical content specification. . . ."

Paragraph 3.1.6.2 says: "Camera-ready copy. Camera-ready copy shall consist of all text pages including tabular data and emergency page markings when applicable and artwork suitable for use in the development of printing plates."

Paragraph 3.3.2 says: "Grammatical person and mood. The second person imperative mood shall be used for procedures ("Remove test set from carrying case.") Third person indicative mood shall be used for description and discussion ("When switch A is in the ON position, lamp 14 lights.")."

A spec *on* technical writing, then, is a set of requirements for writing and producing certain technical publications that are often furnished to a customer with the equipment, as part of the contract. Such specs stipulate arrangement of content, terminology to be used, format expected, editorial and printing requirements, preparation of art work, and other publishing conditions.

GENERAL AND DETAILED SPECS

Frequently, one spec will state a *general* requirement, and an associated spec will *detail* this requirement. This difference is shown in MIL-H-6738A, "Handbooks: Overhaul Instructions with Parts Breakdown (for relatively simple accessories and equipment)." Paragraph 3.5.3, on "Cleaning," tells us to include "the Government specification number of the cleaning agents required." In other words, when we write such a handbook, we must give this specific cleaning information. We turn to such a handbook, say "Overhaul Instructions with Parts Breakdown, Hydraulic Cylinder, Inboard Landing Flap, 542888, Northrop," and we find in the Overhaul Instructions, paragraph 3: "a. Wash with cleaning solvent, Federal Specification No. P-S-661 and dry with compressed air."

Thus we see that some technical writer working for Northrop guided himself by following MIL-H-6738A in writing this overhaul handbook; then he had to learn, from some official source, how to fill in all the information holes indicated by the general spec. He had to discover that No. P-S-661 was the correct type of fluid for cleaning the Northrop Inboard Landing Flap Hydraulic Cylinder.

Often, then, in the writing of technical publications (also, in dealing solely with hardware specs), two further types of specs must be followed, general and detailed. *Many* detailed specs can be written *to* or under a single general spec. Fortunately, federal specs always list, under the heading of "Applicable Documents," any other specs that pertain to the building or writing to be done.

A detailed spec will sometimes—say, in an art requirement—contradict its general spec. This is always a dismaying discovery. Any such clash should be brought to the supervisor's attention and thrashed out.

So our spec picture is becoming confusing. Hardware specs include both general and detailed specs; and so do the specs on producing writing such as handbooks.

SPECS CLASSIFIED BY SOURCE

Specs are also classified into *government, commercial,* and *company* specs. We will consider only specs on writing.

1. The government specs set forth the requirements of various U.S. Government branches for handbooks to be furnished with the material, services, and equipment purchased from private agencies—equipment such as tanks, airplanes, electronic devices, and missiles.

2. The commercial specs are prepared by a company or by an industry-wide association of companies to set up standards for writing the handbooks furnished with the equipment that these companies purchase. The companies' suppliers must write handbooks conformable to such specs.

3. The company spec is prepared by an individual company to standardize and regulate the technical publications that are written by its own personnel, either for customers or for use by employees of the company.

Company specs for technical publications are sometimes part of the company's standard manual of operating procedures, or they may be prepared and used exclusively by the publications group.

Of these three types, the company spec may interest the reader most. In general, the format of company specs is like that set up years ago by the federal government. The student should know what this is so that he can watch for it, and for deviations from it. There may be many such deviations, however, depending on the purpose of the spec. The usual first-order headings are Scope, Applicable Documents, Requirements, Inspection and Tests, Notes and Data. Some government specs may rephrase the fourth heading as Quality Assurance Provisions; instead of Notes and Data, you might find "Preparation for Delivery." There is considerable variation.

"Scope" states the purpose of the spec, its applicability or limitations, and a brief description of the contents. "Applicable Documents" contains an itemized list of documents often referred to in the body of the spec; their exact titles and numbers are given. They are frequently detailed specs. Familiarity with these also-governing specs is assumed. (This means you should get copies, read them, and keep them at hand for ready reference if needed.) "Requirements" is usually the lengthiest and most detailed section because it comprises the basic orders for preparing whatever publication is the subject of the spec. A spec for machine shop handbooks, in this section, would stipulate what plus or minus tolerances were to be included to insure the correct fitting together of parts. Here, for example, are requirements for Section 3 of one company's "equipment specifications."

"Prepare Section 3, using any or all of the following points as they are relevant to the requirement at hand.

"Specify envelope dimensions, bolting circles, dowel pin locations, mounting provisions, and weight limitations. Specify materials to be used, service requirements, fabrication techniques, lubrication, color, finish, and accessories. Specify service life, intermittent or continuous duty, and extreme environmental working and storage conditions. Specify inputs and outputs; flows, pressures, voltages, horsepower, rpm, cycles per second, ranges, adjustments, and tolerances acceptable. Specify type of nameplate and information required. The specification may contain explanatory figures, tables, and charts. Specify packaging and shipping requirements; i.e., the quantity to be packed together, labeling of container, and the degree of protection from the elements, port closures, and protective covers."

Here we have directions for writing a certain part of a company spec, the aim of the spec being to lay down conditions either (1) for manufacturing items within the plant or (2) for contractors who make items that the company plans to buy.

FINAL THOUGHTS

Specs are part of a chain of activity. For example, if Mr. Greenbax, the customer, ordered a fleet of cargo airplanes from the great Zoom Airframe Company, he would be somewhat annoyed to receive fighter planes instead. Also, if Mr. G. specified certain information and illustrations to appear in the handbooks furnished *with* his planes, he would be unhappy if the handbooks did not conform, because his personnel could not use the handbooks as guides when flying, maintaining, and repairing the planes. Mr. G. would probably sue Zoom Airframe Company for violation of contract and mental cruelty. By all means, let's give Mr. G. and his colleagues the kind of handbooks they ask for and need. Such handbooks are "written to spec" by the contractor. Such specs, therefore, *make possible* and *improve* the technical writing that *produces* handbooks that *keep* hardware operating.

But the world of specs is a confusing one. First, as noted, there are so many, of such various types. Second, their importance is sometimes deliberately and confusingly overemphasized. We must warn the reader against overestimating the role of the spec. In the variegated empire of technical writing, certain jobs *do* depend utterly on writing to spec. But there are also many writing and editing jobs—perhaps the majority—to which specs do not apply and for which specs are not used or even mentioned from one year to the next.

Reports, brochures, and other institutional literature, industrial motion pictures, company advertising, revisions and editings in general, and many manuals—these, as such, are not controlled by written specs. Schools that teach writing to spec intensively and expensively as though this were the be-all and end-all of technical writing are falsifying the picture. Preferable objectives are to be able to write well and clearly and, as an aid to versatility and adaptability, to be aware of the major facets of a vast subject.

Sometimes interpretation of an unfamiliar spec is made easier by studying a publication already prepared according to this spec. Thus a writer gets an idea of how the finished publication will look, or—if mistakes have been made—how it should not look.

As requirements and conditions change, specs of all types are often revised or supplemented, sometimes by page additions. A spec that was effective for handbooks telling how to operate Model X may be supplemented or completely superseded when the handbooks for Model Y are prepared the following year. When writing according to spec, one should be sure to have the applicable spec(s) *plus* any permissible deviations or revisions.

Some technical writers have never learned, oddly enough, that certain printed specs control their work. Other writers, disliking specs, never bother to master the relevant ones; consequently, never quite know what they are talking about.

But the employee who *needs* to write from spec cheerfully masters the ones that concern his bread and butter. He knows a relevant spec is as important a tool as his typewriter. He secures personal copies of the specs needed, familiarizes himself with them, reading them all the way through at least once, and keeps them handy for reference. The conscientious writer owes it to his supervisor, his firm, and himself to be sure he understands each applicable spec and knows who the final authority is in interpreting it.

Use of the spec has many advantages. It can provide a ready-made *outline* to follow, hence a checklist to show one's progress in writing it. It may indicate the *level of readers* for whom one is writing, and it often shows how to process *illustrations*. In addition, it gives the *terminology* to be used, establishes *authority* in case of doubt or controversy, adds to professional *skills,* and trains and disciplines the writer.

ON WRITING SPECS THEMSELVES

Every spec that is followed, of whatever type, must have been written by someone. Instead of building or writing to spec, an employee may be asked to write a specification itself, whether about hardware or about a written product. How can he gain the necessary knowledge?

He should first clarify for himself the purpose of the spec. Will it show how to write manuals for his company's customers to use? Will it supply information to his own company's purchasing department as to what is needed to fabricate a certain product and just how to do it? What department is the spec supposed to help—Receiving, Inspection, Tool Planning, Design Engineering, Fabrication? Who is responsible for furnishing the information that is to be in the spec: the standards department, quality control, the purchasing agents, certain engineers? When these questions are answered, the writer-editor can start collecting his material, either by interview or by requesting written contributions.

The special requirements of one's own company must be gathered from within it. But much general, pertinent information may be secured from outside guides and sources such as the American Standards Association, various technical consultants, certain outstanding suppliers or vendors for one's own company, sectional committees from industry, and especially from similar specs. Spec writers should be expert on all such research.

One may decide to use the well-known five-part format for writing specs:

Scope, Applicable Documents, Requirements, Inspection and Tests (or Quality Assurance Provisions), and Notes and Data (or just Notes). Sometimes Preparation for Delivery is also a heading, as we noted.

If one adapts such a general outline to his special needs, the resultant spec will be unique. Of course if a company is just beginning to write its own specs, the Applicable Documents may be few. Often, the Notes section seems to include merely left-over pertinent data; therefore, an ideal might be to write one's spec so well that Notes are nil.

By far the longest of the five standard spec parts is Requirements. These designate the character or quality of materials; formula, design, construction; dimensions, color, weight, nameplates, markings; chemical, electrical, or physical requirements; product and performance characteristics. The Requirements section indicates the general standard of quality and workmanship that a commodity or service must meet to be acceptable. Specific tests are put in the fourth section, on Inspection.

The writing of specs entails important style requirements; strict adherence to them promotes clarity of meaning. The following are a few such requirements.

1. Each frequently appearing phrase should be used in an identical way; for example, "unless otherwise specified" should always appear at the beginning of sentences and, if possible, of paragraphs.

2. Class terms such as "figure," "paragraph," and "bulletin," are capitalized only when they precede numbers, as "Figure 12."

3. Cross-references need be made only to the *number* of other paragraphs.

4. Each major paragraph in a spec should be given an underlined heading for easy reference.

5. The use of symbols, especially those formed of single characters such as ' for foot and x for times or by, should be minimized; a single textual error can destroy the meaning. Such symbols should be expressed in letters of the alphabet (as "percent", not "%"), even though abbreviated (as "ft" for "foot," and "in." for "inch").

6. "Shall" is used to express binding requirements; "will," "should," and "may" are used for nonmandatory directions or suggestions.

The list of style requirements should be fully determined before textual writing is begun. This list can be formed by (1) studying specs and (2) incorporating the style stipulations of one's own company.

A spec dealing with writing itself will include such topics as abbreviations, appendixes, arabic numerals, bibliography, English usage and grammar, equation writing, format, headings, illustrations, makeup, numbered items (e.g., appendixes, equations, headings, illustrations, tables), pagination, reductions in size, references, symbols and units, typography, and tables. In addition, subjects closely associated with editing (such as repro typing and layout) and technical illustrating may have to have directions written for them in the needed spec.

Thus, topic by topic, one builds up the spec that he is writing. One's research requires the reading of many other, usually federal, specs such as the "NASA Publications Manual" (NASA SP-7013) of the National Aeronautics and Space Administration, as well as NASA SP-7007 and 7008, both on spec-writing procedures.

But suppose that what is found does not express one's purpose. For example, under pagination, an already existing spec from which one is "copying" may say that pages are to be numbered with arabic numerals placed at the bottom in the middle of cach text page. But one may prefer that such numbers be placed at the lower left of each left-hand page and at the lower right of each right-hand page. In this event, one simply writes up the page-numbering specification as he wishes it to be done, rejecting the method of the preexisting spec. This spec has at least introduced the problem to the spec writer, giving one definite solution, and suggesting another. The writer, in the last analysis, uses his own judgment.

A well-written spec is meticulously precise in its descriptions and directions. Everything is stated so it cannot be reasonably argued about. No alternatives arc left open to confuse and bedevil a conscientious reader; thus, no part of the spec can be honestly misinterpreted. All applicable, essential matters are stated unequivocally.

He who writes a good spec is, of course, a first class writer, since he must begin by gathering exact data. He must then achieve clarity and economy. He need not worry about emphasis if he has written carefully to the right outline. He must, then, be accurate, clear, and economical. For an inaccurate spec is a crime against knowledge; an unclear spec, an affront to the intelligence; and a wordy spec, a waste of company time.

SUGGESTED FINAL FOR CHAPTER V

Answer, then see Appendix, p. 210. Writing to spec as well as writing a spec itself requires that words be used with the utmost care. This stipulation has already been studied in "How To Write Effective Sentences."

1. Show how specs come to be needed, especially in a new scientific field.
2. Classify specs as briefly as possible.
3. In writing to spec, what do you think would be the greatest difficulty?
4. Explain the difference between writing to spec and writing a spec.
5. State the first three or four things you would do if your company asked you to write a spec on a certain subject.

Appendix

DIVISIONS OF TECHNICAL WRITING

To see that the vast field of technical writing does divide conveniently into the two great types—instructions and reports—let us imagine that we are walking from one department to another in a large plant. We overhear various people engaged in the following discussions and topics of discussion.

1. "Personnel wants us to produce a really attractive recruiting brochure to send to all college men who write in about employment. Do you think you could work up a rough draft in a couple of days, Joe?" This would be in the nature of a *report* to collegians on career opportunities existing at the plant.

2. A request to the main writing group from top management for a style book that would standardize the company's use of titles, symbols, abbreviations, capitalization, nomenclature, and punctuation. Such a book would consist of *instructions*.

3. From a company officer: "Tom, I just don't have time to go to the electronics convention. I know it's important. Would you sit in for me and take notes and then write them up?" *Report.*

4. A planning conference is meeting to initiate the revision of a company familiarization booklet to hand to new employees. *Instructions.*

5. Interviews are being planned for an article on the tool design department, for publication in the company's weekly newspaper. *Report.*

6. "On this 'story board' you've done for our 'Engineering Services' film, Mr. Preston thinks the layouts are great, and the copy too, except for some of the dialogue. He says, could we make it a little more 'institutional'?" *Report.*

7. The placing of want ads to secure new employees would combine *reports* on the company's needs with *instructions* on how interested persons should apply.

8. References to poorly written interdepartmental correspondence which is usually put out in quantity, as a matter of routine. *Reports* mostly.

9. "We've just been assigned a big project, the writing of an 'Engineering Design Manual,' to standardize drawing, drafting, and operational procedures. We've been given a tentative deadline of three months for it; but judging from everything they say they want in it, we'll need closer to six months before it's out." *Instructions.*

10. A conference on writing the vitally important "Annual Report," containing the company's financial statement. *Report.*

11. The scheduling department is checking on the deadline for completion of an operation and maintenance manual, to be furnished with the company's newest product. The manual must contain information on all the latest engineering changes, plus the appropriate illustrations, and must be printed and bound in time to be shipped with the product. *Instructions.*

12. A vice president is working with his aides on a speech he will make to department heads announcing changes in company policies and procedures. *Instructions.*

13. The company's training school has received requests from various departments to present these new courses: "Shop Mathematics," "Etched Circuitry," "Principles of Accounting," and "Quality Control." A technical writer is asked to help the four designated instructors prepare "Course Outlines," then "Lesson Outlines," and edit all these *instructions.* But the technical writer's first task is to prepare a *report* to the training school manager, showing the coverage of subject matter in each of the courses.

14. The personnel manager asks a department head to have five job descriptions written up, covering duties and qualifications of new positions. These job descriptions—hiring *instructions* to personnel—are needed to help in advertising for and interviewing applicants for the new positions.

15. "Bill, the boss wants you to write up the Weekly Activity Report while Hank's on vacation. The Old Man really reads these, so be sure you get all the facts." *Report.*

16. "Now that we're reorganizing the export department, we'll need several new procedures for shipping to foreign countries. Will you meet with the head of export as soon as possible and ask him to let you work with someone who can give you the information you'll need?" *Instructions.*

17. A new executive is requesting some Managerial Reports, a regular research service available to department heads who need special information. *Reports.*

18. A conference is called to rewrite and clarify a dozen confusing passages in the large book of company policies. *Instructions.*

19. "We'd like you to write the biweekly leaflet comprising both safety and conservation practices." *Instructions.*

20. "Your work will be to write what we call 'Justifying Letters' to the purchasing department, explaining why we need various new equipment. If these letters aren't strong, we find that purchasing pays no attention to us. We've had a high batting average lately by pointing out what will happen if we *don't* get the item. Your first request should be for an IBM (International Business Machine)." Such letters combine *reports* with *instructions.*

ANSWERS TO EXERCISES 1–6

Answers to Exercise 1, p. 10

1. Anyone planning a technical career should be able to write well because he will probably be called on, sooner or later, to do exactly that.

2. The reasons given for technical writing being so different from the usual college English course are that (a) it is understood that instruction in technical writing will pay off in an assignable vocational future, whereas in the latter case of college English no one believes it will; (b) technical writing involves exposition and description, whereas college-theme writing stresses narration and argumentation; (c) control in technical writing is objective facts-to-be-found rather than theme choice plus a creative literary imagination.

3. The best preparation for technical writing fosters a general competence in the field because exactly what branches of it one will find himself in, and just what kinds of writing products will be required of him, are not known while he is still a student.

4. There are no sciences or techniques which cannot benefit by being explained in clear technical English. If trainees need to be taught any science or technique by reading about it, should not what they read be skillfully written? Today, certainly, word-of-mouth training is inadequate for mass instruction in the sciences and technologies.

5. The most difficult task for a technical writer is probably to plan for writing; the writing itself tends to follow easily from a good plan. The topics in the outline, which is the plan, can serve as subjects for sentences; the outline is the skeleton for the final product.

6. All faulty plans, applied, result in faulty performances. Some outlines are faulty plans, applied. Therefore some outlines result in faulty performances.

7. The two main reasons for failing to include enough in a writing plan are not thinking hard enough about one's subject and neglecting to balance one's outline by thinking parallel thoughts.

8. "Machine shop theory and practice." As the text points out, heading "III. The Lathe" fails to have, consistently, "A. Definitions" under it. The same applies to "Milling Machines."

"Industrial engineering." "Material loss" should have been an item in dealing with the 55-gallon freon drums.

"The gasoline engine." Two headings, "D. Exhaust" and "B. 3. Ignition of mixture by spark," and possibly others, have been left out.

"Printing." "Time Required to Print" should have been somewhere in the outline.

"Magnetism." "How Magnetism is Measured," and possibly other topics, were left out.

9, 10. Answers to these questions are based on the student's own experience, rather than on the text.

Answers to Exercise 2, p. 15

1. Three reasons were given to explain why we let irrelevant topics creep into our writing plans: mistaken enthusiasm, illogical analysis, and a narrowing of our original topic.

2. The student supplies this answer.

3. When one allows an irrelevant topic, or several, to remain in his outline, they become, when written up, useless "clinker" paragraphs that confuse and irritate the reader.

4. The best way to exclude needless topics from an outline is to make sure that each topic retained is included under or belongs to a limiting sentence.

5. One should use the limiting sentence by referring to it as each topic is examined. A separate act of judgment should be made as to whether each topic is or is not a part of the subject expressed by the limiting sentence. No topic that is not such a part should be retained in one's outline.

6. One sometimes needs to rephrase or amend a limiting sentence because research has turned up certain facts so important to the subject that they change one's purpose. An instance is given in the text (p. 14).

7. To say the limiting sentence is an ideal in technical writing indicates that it should be used when possible, but sometimes cannot be. To review the reasons why this is so, please reread p. 14 on the lack of a limiting sentence.

8. A limiting sentence could not only help one organize a report, but could appear somewhere in it, perhaps several times, and be expressed in different ways.. It might be the major conclusion. Examples are the limiting sentences on p. 13.

9. A textbook would not likely be written around a limiting sentence because outlines dealing with complex subjects are usually not expressible in any helpful key sentences. What one limiting sentence, for example, could sum up a textbook on physics, geology, electronics? Thousands of *sub*topic sentences, so to speak, make up such a subject—one too comprehensive for a single predicate. A text is not a yes-or-no situation, but a compilation of a host of tried and true facts. Limiting sentences, on the other hand, are ideal for reports that express particular points of view.

10. A subject guides the outliner better if it has a predicate simply because a statement that takes a precise point of view is thereby formed. This narrowing of the field of interest makes it easier for the outliner to choose only those topics that support the chosen statement.

Answers to Exercise 3, p. 25

1. Saying that vertical and horizontal movements of thought tend to alternate in outlining refers to the fact that one begins by dividing a topic into its constituent topics; this act is a dropping down to the next lower level of heading. One continues by carefully scrutinizing these constituent topics or coordinates to make sure they belong together—a kind of horizontal check. Then, if there is more to the outline, each of these first-order headings is examined to see if it divides into second-order headings. If so, this is the second vertical movement of thought; and so on, until the outlining is completed.

2. Correlatives are the "close relatives" that each topic in an outline has, one such being the superordinate from which it stems, and the other(s) being at least one coordinate resulting from the division of the superordinate. These coordinates help to make outlining easier because we know they must be there and because they indicate one another. Any subordinate has to have a superordinate and at least one coordinate.

3. For two or more topics to be coordinates, they must have the same superordinate—that topic of the next higher order from whose division they resulted.

4. Yes, these two statements are true. Any topic and its subordinate have meaning when considered exclusively together; there is no relationship as close as this. For one topic is a part of the other and therefore subordinate to it. But if, to relate two topics meaningfully, a third topic is needed, it indicates that the two topics do not have that close relationship of one being a part of the other; rather, both are parts of this superordinate.

5. The two types of logical acts that we keep making in outlining are subordination and coordination. The first type requires three topics, namely the superordinate and at least two subordinates. The second type requires three topics for existence, namely at least two coordinates and the superordinate whose division has made them coordinates.

6. An informal outline is one that evolves in the course of research and/or in adopting certain major headings formally required by the type of writing one is planning. The informal outline is usually retained in memory only, not written out, not deliberately worked upon, and never seen. It is inadequate because it robs us of the chance to visualize our outlining decisions and to correct the incorrect ones.

7. Failure to subordinate and coordinate topics correctly is more upsetting to the reader than is failure either to include every topic required by the subject or to exclude every topic not required by the subject. This is so probably because infractions of Rules 1 and 2 still leave the reader with the nucleus of a well-outlined subject, whereas an infraction of Rule 3 destroys that nucleus and basis.

8. When we "got stuck" in the course of outlining "Heat," we first confirmed the progress we had made, then verbalized what the next step was. We had to discover that one or more of the first-order headings we were already sure of was a superordinate for one or more of the three topics that we had not positioned as yet. In other words, in order to proceed we had to grasp clearly what the next act of division was.

9. Trying to use a solution to solve more than one problem means, usually, trying to find more than two subordinates. We do not stop there, but ask ourselves, "Are there three, four . . . ?" We keep using our original reason for dividing the topic into the two subtopics or subordinates because this reason may apply to still more subtopics.

Or suppose, looking at this same situation differently, the outliner discovers a subordinate's fellow subordinate. He does not stop there, but asks, "Is there a third of these coordinates, a fourth . . . ?" It may be easier for him, in certain cases, to think of these topics as coordinates than as subordinates. If so, he takes the easier path psychologically, knowing that the result is the same. But whatever he is doing, he finishes. While he's at it, he extends his search over the entire set of similar topics with which he is dealing in order to maximize his outlining discoveries.

This practice fulfills the "all" part of Rule 3, whether we look at it from the vertical or horizontal point of view. That is, finding all of a topic's subordinates is exactly the same thing as finding all of any one of those subordinates' coordinates.

10. Each topic, when correctly placed in an outline, always has at least two "close relatives" because it requires not only the topic that "fathered" it, of which it is a part, but also at least one "sibling" topic that would be left over from dividing the superordinate parent.

11. There must be at least two coordinates subordinated to a subject when it is divided because dividing anything makes at least two parts.

Only *if* a topic is divisible, must it be divided into at least two subordinates. All topics at the bottom of an outline, so to speak—its lowest subdivided level—must be, if not indivisible, at least undivided for our purposes; otherwise, dividing would go on forever, and no outline would ever be completed.

12. The reader's answer suffices.

13. Every topic in an outline (except the subject) requires both a subordinating and a coordinating decision. The former decision designates that superordinate of which the topic is a part; the latter decision designates the at-least-one-other topic that is its equal.

14. In saying that "outlining progresses by our testing one small theory after the other," one refers to the fact that he proceeds, at any one time, to prove or disprove a certain limited viewpoint about the outline. These theories or viewpoints are guides for construction. When one is proved or disproved another is formed, referring to the next bit of work to be done.

These theories involve both subordinating and coordinating, and could be of the following types: "Topic A must have subordinates that I can find because Topic B has none; B is the only coordinate of A, and there are still topics to be classified." Or, "Probably among eight topics still unexamined, I can find more coordinates of Topic M." Or, "There are so many topics to classify that probably several orders of headings are involved."

15. A heading is a topic that functions in an outline or in a writing and is therefore usually described as to importance and place by a preceding symbol such as "II," "A," "6," and so forth.

Outlining is a kind of classifying done characteristically for writing products. In the interests of science and precise thinking, one classifies various types of objective facts such as the one million biological species; one *outlines* the topics in something he has chosen to write about.

Answers to Exercise 4, p. 28

1. Yes, Rule 4 is independent of the first three rules given for outlining. You could include every topic required to outline a subject, and exclude every topic not required, and still have the topics disordered; and they could be out of order even though they were all subordinated and coordinated correctly.

2. Two aids in applying Rule 4, that is, ordering topics of an outline correctly, are that (a) the research we have done will already have been ordered to some extent and (b) our company instructions will probably be to write in the form of a model already available to us. A third possible aid is that when "writing to spec" we will already be following a basic outline.

3. The chronological type of outline helps one to apply Rule 4 because the time sequence itself constitutes an ordering of topics; you place the initial event first, the next one that happened second, and so on to the last event in time.

4. If an entire outline is chronological in type, its lower-order headings would also tend to be chronological. If major events are expressed in a time order, that is, from first to last, or vice versa, the minor events composing it would also tend to be arranged in such a time sequence. This, however, need not always be the situation. A minor event could be presented according to a different principle, as by first giving a result, then tracing it back to its cause. Such reversing of the time order and other variations in the ordering principle might be effective in places.

5. Technical writing so frequently involves the chronological type of outline because both reports and instructions, the two major types of such writing, involve references to time. Reports, being concerned with past happenings, are usually most clearly given by letting time do the ordering. Instructions, concerned as they are with what persons should do, involve the listing of steps that are to be taken in an orderly time sequence.

Answers to Exercise 5, p. 35

    I. First-order Heading
      A. Second-order Heading
        1. Third-order Heading
        2. Third-order Heading
          a. Fourth-order Heading
          b. Fourth-order Heading
          c. Fourth-order Heading
        3. Third-order Heading
        4. Third-order Heading
      B. Second-order Heading
    II. First-order Heading

Answers to Exercise 6, p. 35

1. The symbol for a fourth-order heading, in the usual format as shown on p. 34, is a small letter of the alphabet. The seventh such heading would be the seventh letter of the alphabet, or g.

2. C.

3. A seventh-order heading was said on p. 34 to be a small roman numeral. The fourth such number would be iv.

4. A.

5. 2.

6. (c)

7. A fourth-order heading, a small letter, will be used, starting with a, as often as it is needed, as many times as there are headings of this type to be represented. This will depend upon the particular outline that is being made, so the question cannot be answered in general.

8. There could not be an "only" third-order heading, since there must be at least two headings of each type in an outline.

PAGE FROM GPO MANUAL

Page 1 of the GPO: U.S. Government Printing Office Style Manual, 1973

1. Suggestions to Authors and Editors

1.1. This STYLE MANUAL is intended to facilitate Government printing. Careful observance of the following suggestions will aid in expediting publication and in reducing printing expenditures.

1.2. Copy must be carefully edited in accordance with the style laid down herein before being sent to the Government Printing Office. Changes on proofs add greatly to the expense and delay the work.

1.3. Legible copy, not faint carbon copies, must be furnished. This is essential in foreign-language copy and in copy containing figures.

1.4. Copy should be sent flat, with the sheets numbered consecutively, and typewritten on one side of the paper only. If both sides of reprint copy are to be used, a duplicate must be furnished.

☆ 1.5. To avoid unnecessary expense, mutilation of copy, and to expedite GPO production, each page should begin with a paragraph.

1.6. Tabular matter and illustrations should be on sheets separate from the text, as each is handled separately during typesetting.

1.7. Proper names, signatures, figures, foreign words, and technical terms should be written plainly.

☆ An open-face star preceding a paragraph indicates that a substantial change has been made in the wording or meaning of that rule. A star does not appear where a simple renumbering of rules has occurred.

1.8. Manuscript and typewritten copy in a foreign language should be marked accurately as to capitalization, punctuation, accents, etc.

1.9. Footnote reference marks in text and tables should be arranged consecutively from left to right across each page.

1.10. Photographs, drawings, etc., for illustrations should accompany the manuscript, each bearing the name of the publication in which it is to be inserted and the figure or plate number. The proper place for each text figure should be indicated on the copy by inserting its number and title. If the legends are placed on one or two sheets of the manuscript copy, it is preferable that the copy for the legends be placed at the beginning of the manuscript to facilitate the placing of the legends in the proper position.

1.11. A requisition for work containing illustrations must be accompanied by a letter certifying that the illustrations are necessary and relate entirely to the transaction of public business (U.S.C., title 44, sec. 118). The total number of illustrations and the processes of reproduction desired should also be indicated. Instructions should be given on the margin of each illustration if enlargement or reduction is necessary.

1.12. If a publication is composed of several parts, a scheme of the desired arrangement must accompany the first installment of copy.

☆ **1.13.** To reduce the possibility of costly blank pages, avoid use of new odd pages and halftitles whenever possible. Generally these refinements should be limited to quality bookwork. (See rule 2.85.)

ANSWERS TO EXERCISES 7–17

Answers to Exercise 7, p. 40

1. The difference in structure between a topic and a sentence outline is that the first is an ordered system of subjects only, whereas the second is an ordered system of statements made about those subjects.

2. It is possible to outline what one is going to write about, but not what he is going to say, by making a topic instead of a sentence outline. In a topic outline, no statements are made. Only by inference one is saying, "In my final written product, I am going to make statements about all these topics."

3. The topic outline does not tell where Atlantis was believed to be, but the sentence outline does. This basic difference between the two types of outlines is explained by the answer to Question 1.

4. The things that must be done to change a topic outline into a sentence outline are various. Please reread the gasoline engine example beginning on p. 38. Of I, true, a simple statement was made. But the topic of I.A, "Intake," became only an adjective in the sentence made from it! The

MIL-P-38790
1 September 1968

SUPERSEDING
(Refer to 6.3)

MILITARY SPECIFICATION

PRINTING PRODUCTION OF TECHNICAL MANUALS: GENERAL REQUIREMENTS FOR

(Photolithographic Negatives, Reproduction Assembly Sheets, Printing)

This specification is mandatory for use by all Departments and Agencies of the Department of Defense.

1. SCOPE

1.1 This specification covers the requirements for photolithographic negatives, reproduction assembly sheets, and printing of the type of basic technical manuals indicated, including changes, revisions, supplements, and reprints thereto, and thereof.

1.2 Certain paragraphs in this specification are not applicable to all Services. The paragraph text is prefixed with the letter (A) Army, (N) Navy, (MC) Marine Corps, (F) Air Force, as applicable. The word "All" indicates that the paragraph is applicable to all Services.

2. APPLICABLE DOCUMENTS

2.1 *Government documents.* The following documents of the issue in effect on date of invitation for bids or request for proposal, form a part of this specification to the extent specified herein.

SPECIFICATIONS

Military

MIL-M-38784   Manuals, Technical: General Requirements for Preparation of

PUBLICATIONS

Department of Defense

DOD 5220.22-M   Industrial Security Manual for Safeguarding Classified Information

(Copies of documents required by contractors in connection with specific procurement functions should be obtained from the procuring activity or as directed by the contracting officer).

Joint Committee on Printing, Congress of the United States

No Number   Government Paper Specifications Standard

Application for copies should be addressed to the Superintendent of Documents, U.S. Government Printing Office, Washington, D. C. 20402.)

2.2 *Other publications.* The following documents form a part of this specification to the extent specified herein. Unless otherwise indicated, the latest issue in effect on date of invitation for bids or request for proposal shall apply.

ASA PH 1.25-1956 Safety Photographic Film

(Application for copies should be addressed to the American Standards Association, Inc, 10 East 40th St., New York, New York 10016.)

3. REQUIREMENTS

3.1 *Photolithographic negatives.* (All) Negatives shall be of first class standard photolithographic film and furnished in such condition that no additional work will be required on them prior to printing. Negatives shall have a film base (cellulose acetate, triacetate, polystyrene, polyester, vinyl, etc.) and shall conform to American Standards Association specification for Safety Photographic Film, ASA PH 1.25-1956. Thin base film (less than .004 inch in thickness—refer to 3.1.4 for exception) and paper base material are unacceptable. Film with a slight matte finish on the emulsion side is acceptable. Tests will be conducted to determine compliance with the requirements herein. The contractor will be required to remake negatives deemed unacceptable by the procuring activity at no additional cost to the Government.

3.1.1 *Negative dimensions.* (All) For layout purposes, negatives shall be trimmed so that regardless of length, the top margin shall always be 3/8 inch, plus or minus 1/32 inch; regardless of width, the outside margin (side away from the binding edge) shall always be 3/8 inch, plus or minus 1/32 inch. Therefore, in no case shall the margin from the printing area to the negative edge be less than 11/32 inch. The dimensions of single page negatives shall be uniform within a manual.

3.1.1.1 *Other sizes.* (All) When circumstances require that a Service procurement manual of a size different than cited in the following chart, the dimensions of the negatives shall be as specified in the contract.

noun in the I.A.1 topic, "Increase," was changed into a verb! So was I.A.2, "Decrease." In the case of I.A.3, "Entrance" became a verb; so did "Closing" in I.A.4. In each case, the writer has to think of the meaning he wishes to create by making the sentence.

5. Your answer suffices.

6. We probably use topic outlines because we never realized that sentence outlines can be an improvement. When possible, they should be used.

7. Three reasons to outline by sentences rather than topics are that this practice (a) clarifies our subject, thus making it easier to write, (b) makes our purpose much easier to grasp by a superior, and (c) composes important sentences for us to use later.

8. It would be preferable to prepare a topic outline rather than a sentence outline whenever one's purpose is not to make debate choices and defend positive theses, but merely to discuss alternatives or situations, leaving it to others to come to final conclusions. This answer probably does not explain why there are so many more topic outlines than sentence outlines. We suspect it is because topic outlines are easier.

9. The difference between a writing outline and a reading outline—that is, one formed to write from and one formed so as to review and learn what one has read—is as follows: The first type proceeds from outline to writing and is usually topical, whereas the second type proceeds from reading to outline and is always made up of sentences rather than just topics.

10. In terms of successfully making learning outlines, the difference between efficient and inefficient learners is that the former *continues* to subordinate and coordinate his various cards (with the sentences on them) until he has *finished* the outlining function and has a schema of the relative importances and relationships of all the sentences. The inefficient learner stops in mid-course, without finishing the outline, and thus fails to see and learn the subject as he should and as the author probably wanted him to.

Answers to Exercise 8, p. 52

1. He resigned during the experiment, but was asked to stay until it was completed.

Original: "During the course of the experiment, he resigned; but he was asked to stay on until it was completed." Contributing nothing, "course of the" was left out. "During the experiment" was placed after "he resigned," so as to make the "it" reference clearer. "On" was deleted because "to stay on" is poor usage. The punctuation was also simplified, from comma and semicolon to just a comma. All such changes help readers to understand. The conscientious writer is willing to take pains to make these slight improvements. The overall effect is worth the trouble.

2. To permit operational checkout of the handling equipment, a test installation was constructed.

3. Since I know the check will bounce, endorsing it is unnecessary.

Original: "Since I know the check will bounce anyway, the customary practice of endorsing it on the back is entirely unnecessary." Notice that "anyway," "the customary practice of," "on the back," and "entirely" have been dropped. Such expressions are unnecessary.

4. He remained at his old post.

Original: "He continued to remain at his old post." Cutting out "continued to" may be debatable. At first, one may feel he is losing values by such deletions. A closer inspection usually reveals that the values are too slight to save; or, merely, that our characteristic wordiness is to blame.

5. To compress crimping dies, engage foot lever, which, when down, will leave dies in closed position.

Original: "Engage foot lever to compress crimping dies; then, when foot lever is down, the crimping dies will be in a closed position." Repetition of "foot lever" has been avoided by placing the term further into the sentence so that "which" can be used instead. Other changes have been made; this sentence required rewriting.

Answers to Exercise 9, p. 53

1. Good weather prevailed throughout the flight.

Original: "Good weather conditions prevailed throughout the flight." (Seemingly necessary.) The word "conditions" added nothing.

What subclass of the unnecessary does the next original sentence exemplify? Such a classification appears in parentheses after each original version, as above.

2. Hurry and connect the battery.

Original: "Hurry and connect up the battery." (Poor usage) "Up" was deleted. Several examples of these useless tag ends have already been given. Discovering them is an excellent way for the student to start making a habit of applying Rule 1.

3. In trying to apply the plan, we experienced many difficulties.

Original: "We encountered frustration in applying the plan, and experienced many difficulties." (Requires rewriting.) This is an example of a sentence in which apparent repetition of ideas is noticed first, before mere repetition of words. One suspects that encountering frustration and experiencing difficulties refer to the same situation, which should be expressed only once. One, then, must either second-guess the writer or ask him what he meant to say. By "frustration" did he mean failure, or didn't he? Were they still applying the plan? Yes? Then it hadn't failed yet, and "frustrating" was incorrect, besides being useless. A basic choice of meanings must be made; was the plan defeated, or only complicated by troubles?

One cannot always go to an author and ask what he meant. One can, though, consult a dictionary; students are urged to form this habit.

4. The next installment will discuss radioactivity and its relation to nuclear energy.

Original: "The next installment will discuss radioactivity and its relation to nuclear energy—in a forthcoming issue." (Seemingly necessary.) The next installment would have to be "in a forthcoming issue," a phrase that thus becomes unnecessary.

5. Completion of the run is scheduled for February.

Original: "Final completion of the run is scheduled for February." (Poor usage.) The "Final" before "completion" was dropped because usage and common sense require that the meaning of a word not be repeated by a word that modifies it.

"Rainy rain" and "strong strength" are obviously silly. There are many other expressions, however, that must be considered carefully in order to detect the repetition of meanings and hence the unnecessary elements in them. Take as an example "*most* unique": since "unique" means being without equal, and the only one of its kind, "most" adds nothing. Or "irregardless": since the suffix "-less" means here "without regard," the prefix "ir-" meaning "not" would falsely signify "*with* regard." Or "true fact": although, strictly speaking, only propositions (statements) are true (or false), whenever a fact is so expressed, it becomes automatically true, otherwise it would not be a fact.

6. The main coolant system conducts the reactor heat.

Original: "The main coolant system is utilized to conduct the reactor heat." (Seemingly necessary.) The deleted "is utilized to" merely takes up space in the original sentence. The verb "conducts" adequately indicates that its subject, "The main coolant system," is engaged in a specific activity. Therefore this fact need not be stated again. Of course "conduct," which was "to conduct" originally, must have an "s" added to make the correct present tense, third person singular form.

7. He ignored the fact that we had not received a work order.

Original: "He completely ignored the fact that we had not received a work order." (Debatable.) This is one of those seemingly borderline cases—until one consults a dictionary. "Ignored" is one of those words with an absolute meaning. Like "necessary" and "perfect," it does not admit of a comparative degree: it means "completely" to begin with. "Ignore" means "refuse to take notice of," and "reject as ungrounded." One does not *partly* refuse to take notice of, or *partly* reject as ungrounded. Since it can be assumed that "he" completely refuses or rejects, "completely" is needless.

8. This procedure established ("proved" would be better) that defects could be removed.

Original: "By this procedure it was established that defects could be removed." (Correct but less effective.) The words "by" and "it was" were deleted from the original. The sentence was thereby strengthened in two ways: it became shorter and was changed from the passive to the active voice.

9. This report summarizes the costs incurred for engineering, construction, and testing.

Original: "The purpose of this report is to summarize the costs incurred for engineering, construction, and testing." (Debatable.) Five words were struck, "The purpose of" and "is to." It can be assumed the reader knows a report has a purpose and that when he reads a textual description of what a report does, he knows the purpose is being stated. This is especially apt to be true if the sentence in question begins an introductory section.

The infinitive "to summarize" was also changed.

10. A thorough chemical analysis (Table 4) was performed on chips taken from the control ingot.

Original: "A thorough chemical analysis was performed on chips taken from the control ingot. This analysis appears in Table 4. (Required rewriting.) Four words, "this analysis appears in," are saved here, merely by enclosing "Table 4" in parentheses; the annoying repetition of "analysis" is also avoided.

This case introduces a new consideration that shows that the rule "use only necessary words" is of wide application and can therefore be of much help to us. The original example consisted of two sentences; it was converted into one. But how can you tell when to make one sentence out of two? Answer: begin to consider the possibility when you notice *repetition*. "Analysis" was repeated in the second sentence, which said only that the analysis was in Table 4. Why waste a sentence to say this when the thought can be reduced in importance by being tucked away parenthetically, as shown.

The student should take pains to develop a keen eye for noting *unnecessary* repetitions in adjacent sentences. He should not be unduly influenced by a separate-sentence situation: many pairs of sentences can profitably be combined.

Answers to Exercise 10, p. 55

1. I remember all my colleagues at the plant better than I do Fred. *Or,* I remember all my colleagues at the plant better than Fred does.

Original: "I remember all my colleagues at the plant better than Fred."

This sentence was equivocal, meaning one or the other of the two corrected sentences above. The choice made depends simply on what one wants to say. Notice that in each case the correction requires that words be added, that is, that the construction be completed. The first correct sentence needed "I do"; the second, "does."

2. The program is broadened from that of last year.

Original: "The program is broadened from last year." The sentence involves a comparison of two programs; not a comparison of a program with a year.

3. Entering the laboratory, we (they, or whatever is required) saw the damage.

Original: "Entering the laboratory, the damage was seen." Since the action of entering the laboratory has no sensible subject to refer to, one must be supplied. It will be "we," "they," "he," "Conningsby Q. Dibble," or whatever the fact of the matter dictates. This supplying of an agent for the action changes the independent clause from the passive voice ("the damage was seen) to the active voice ("we saw the damage.").

4. The results of these tests showed the following. (Or "were as follows.")
   a.
   b.
   c. (and so on)
Original: "The results of these tests showed that
   a.
   b.
   c. (and so on)
What we are correcting is the failure to finish a sentence at the right place. Notice that if the beginning sentence is not concluded before the listing (1, 2, 3, etc.) begins, it becomes part of the single long sentence and should be punctuated accordingly, to the very end of the list. This usage is cumbersome and unnecessary. It is easy to add something like "the following" in order to make a sentence before the listing begins.

5. Encouraged by public approval, the government (for example) stepped up the man-on-the-moon tests.

Original: "Encouraged by public approval, the man-on-the-moon tests were stepped up." This case and 3 are of the same type although a noun, "government," was used here instead of the pronoun "we." There would of course be many ways of writing both these sentences, depending on the truth of the situation: "Entering the laboratory, both men saw the damage," "Encouraged by public approval, the president stepped up the man-on-the-moon tests," and so on.

Answers to Exercise 11, p. 55

1. This last model of sweeper was rejected by every housewife.

Original: "This last model sweeper was rejected by every housewife." A "model sweeper" would indicate an ideal one, and this was not the meaning intended. Also, the prepositions in prepositional phrases should not as a rule be left out.

2. Having placed a guard at each gate, they have lessened the danger of security violations.

Original: "Having placed a guard at each gate, the danger of security violations is lessened." Since the "danger" itself could not have placed a guard at each gate, an adequate human agent was required to perform this function. "They" as well as a host of other nouns and pronouns could serve, depending on the facts.

3. Working overtime every night enhanced his reputation as a company man. *Or:* By working overtime every night, he enhanced his reputation as a company man.

Original: "Working overtime every night, his reputation as a company man was enhanced." Notice that neither of the two corrected versions has the man's reputation itself doing the overtime work. Other correct versions are possible, but each would have to avoid making this mistake.

4. Conclusions reached from the previous investigation included the following.

    a. The weld material showed satisfactory ductility.

Original: "Conclusions reached from the previous investigation included:

    a. The weld material showed satisfactory ductility.

The first sentence should be so written as to end before any list begins. Why include the entire list in the sentence? Besides, the original version was ungrammatical: "Conclusions reached . . . included: The weld material showed . . . ," etc!

5. A good engineer must know how to apply mathematics and how to be practical.

Original: "A good engineer must know how to apply mathematics and be practical." Unless the second "how to" is inserted, it is not clear that the second assertion of the sentence is not merely, "A good engineer must be practical."

6. We advised him that to remain silent was insubordinate but that to refuse outright was worse.

Original: "We advised him that to remain silent was insubordinate but to refuse outright was worse." Addition of the second "that" confirms the fact that two bits of advice were given.

7. The tone of his last talk to us is different from his former talk to us, yet reminiscent.

Original: "The tone of his last talk to us is different yet reminiscent of his former talk." If the last talk was different from his former talk, then the preposition "from" needs to be retained so as to connect the two talks, thus permitting them to be compared.

8. Reading electronics and Sanskrit, he (or whoever was the agent) passed the time quickly.

Original: "The time passed quickly, reading electronics and Sanskrit." Same type of criticism as 2 above; "the time" did not do the reading. "He" or some other agent must have done it.

9. One should not object to summarizing merely on the ground that it is repetitious.

Original: "One should not object to summarizing merely on the ground it is repetitious." The connective "that" is needed.

10. The workbench was cleared, and the models were placed on it, ready for our inspection.

Original: "The workbench was cleared, and the models placed on it, ready for our inspection." If this is treated as an elliptical expression (that is, as one that legitimately leaves out a word or more), we find ourselves assuming that the word is a second "was." In this case, we would be meaning, ungrammatically, "the models was placed on it. . . ." If "models" became singular, the ellipsis would be correct. Thus, "The workbench was cleared, and the model placed on it, ready for our inspection."

Answers to Exercise 12, p. 58

1. A study resulted in selecting three halogens. *Or:* Results of a study were the selection of three halogens.

Original: "A study was carried out, resulting in the selection of three halogens." The important part of this sentence is not the carrying out of a study, yet it is expressed as the only independent clause. The result of this study, the selection, is given the mere status of a phrase. The amended sentence consists of but a single independent clause that ends with a statement of the result. "Carried out" could safely be dropped since the study, if it resulted in the selection, was obviously carried out.

2. The design changes (Figures 1 and 2) stipulate an overall height of 15 ft for the hoist mechanism and cask.

Original: "The design changes are shown in Figures 1 and 2, which stipulated that the hoist mechanism and cask will have an overall height of 15 ft." This sentence shows the same type of mistake as does 1. The minor thought, about where the design changes are shown, is expressed in the independent clause; the major thought, about overall height, is in the dependent clause. In the corrected sentence, what was the entire predicate of the minor thought is tucked away in parentheses. A prepositional phrase, "*for* the hoist mechanism and cask," replaces "*that* the hoist mechanism and cask *will have*." This replacement permits the dependent clause to be done away with so that the entire expression can be integrated into a single independent clause.

3. Sealing capabilities are discussed in this section.

Original: "This section discusses sealing capabilities." Although in general it is preferable to put sentences in the active rather than in the passive voice, there are important exceptions. Here it was decided that, since first impressions tend to be stronger than any others including last ones, the sentence should be started—even though this throws it into the passive voice—with the most important entity.

4. Five copies of progress reports, to be submitted every two weeks throughout the investigation, will give the following information.

Original: "Five copies of progress reports shall be submitted every two weeks throughout the investigation, giving the following information." Notice that this sentence features the fortnightly submittal of reports over the information they are to give. But the frequency of submittal should probably be subordinated to the report information that is desired. So "to be" instead of "shall be" puts the frequency of submittal part in its place, so to speak; "will give" instead of "giving" does the same thing for the report information part. The relative roles of the two "ideas" in the sentence (the "how often" and the "what") have been switched.

5. Today, as never before in history, we need to think out the right relation between science and society.

Original: "Today we need to think out the right relation between science and society, as never before in history." The meaning of this sentence was not changed, nor were there an independent and dependent clause to reevaluate. But a "suspended" sentence was made out of a "loose" one merely by placing "as never before in history" near the beginning. Such improvement can often be made; one should train his "sentence sense" to be sensitive to the possibility.

Answers to Exercise 13, p. 58

1. After the deficiencies were reported, the vehicle was modified.

Original: "Before the vehicle was modified, the deficiencies were reported." Making the reasonable assumptions that reporting of the deficiencies resulted in the vehicle being modified and that this modification is of primary interest, we should not permit the sentence element in which it occurs to be a dependent clause. Therefore we change this clause to an independent one and place it last in the sentence.

2. Although the production of both synthetic gold and diamonds is possible, it is not feasible.

Original: "Production of both synthetic gold and diamonds is possible; however, it is not feasible." The first of two independent clauses was downgraded to a dependent clause, because the concluding clause was more important.

3. Having assembled the supervisors, the manager told them there would be a company reorganization.

Original: "The manager assembled the supervisors and told them there would be a reorganization." The assembling, which was only a means to the telling, was subordinated.

4. Having had to spend an extra month on the plans, they lost the $1000 bonus promised them for a speedy submission.

Original: "They had to spend an extra month on the plans, thus losing the $1000 bonus promised them for a speedy submission." Here the extra month is featured over the bonus loss, a (probably) faulty evaluation that was corrected in the rewrite.

5. Lead pipes contain the coolant.

Original: "Coolant is contained by lead pipes." If, as we assume, it is more important to feature the lead pipes than the coolant, they should be the subject of the sentence. Also, the rewrite in the active rather than the passive voice makes for a stronger sentence. The moral is to make sure one knows exactly what he wants to feature, so that he can compose accordingly.

6. To prevent its collapse, the nickel cell was enclosed in an evacuated porcelain tube.

Original: "The nickel cell was enclosed in an evacuated porcelain tube to prevent collapse of the cell." The author may have tried, once, to avoid repeating "cell" by using "it" as follows: "The nickel cell was enclosed in an evacuated porcelain tube to prevent its collapse." Then he realized that "its" referred grammatically to "tube" rather than to the cell. So he accepted the repetition. But there was a solution: to place the "to prevent . . ." phrase at the beginning. "Its" would then have to refer to the cell. It was not necessary to say "cell" twice (Rule 1). Also, by starting with "to prevent . . . ," emphasis was increased (Rule 3). The sentence was thereby changed from a loose to a suspended one. That is, the independent clause and therefore the affirmation are not finished before the last word of the sentence. The preference for suspended sentences is a habit easy to cultivate and highly profitable style-wise. A loose sentence should be used here and there, however, to prevent monotony.

7. Three scheduled investigations, to be conducted every week, will give the following information.

Original: "Three scheduled investigations shall be conducted every week, giving the following information." That the investigations are to be conducted every week is probably not as important as the information they will present. Therefore the latter should be emphasized.

8. Curves formerly included for thermionic systems were omitted. Part of the curve for thermoelectric systems was dotted.

Original: "The writer omitted formerly included curves for thermionic systems. He dotted part of the curve for thermoelectric systems." In this case the active is not preferable to the passive voice, since it features "the writer" and "he." We are more interested in the curves, which accordingly become the sentence subjects, even though they require passive voice constructions.

9. Additional requirements for withdrawing the material, which neither replace nor contradict those of Procedure G-712, will be found in Procedure G-710.

Original: "Additional requirements for withdrawing the material will be found in Procedure G-710, and the requirements neither replace nor contradict those of Procedure G-712." The two statements joined by "and" are coordinated, whereas the intent was merely to include the G-712 warning while making the new G-710 announcement. Consequently the former was subordinated.

10. A chemical analysis (Table 4) was performed on chips taken from the zirconium control ingot.

Original: "A chemical analysis was performed on chips taken from the zirconium control ingot. This analysis appears in Table 4." Where the analysis appears is given equal status with the much more important first statement. Another fault is the needless repetition of "analysis." So the correcting of these sentences involves dealings with combinations of faults. Accordingly, to make an elliptical expresssion (ellipsis) of "Table 4" not only precludes giving a false status to the second sentence but also saves the words "this analysis appears in." We are preparing to consider combinations of faults. This will introduce no new principles, but will make us warier and less easily satisfied with our sentences.

Answers to Exercise 14, p. 62

1. I told my supervisor only the truth. Or, less colloquially: I told my supervisor the truth only.

Original: "I only told my supervisor the truth." We have already seen how shifting the position of "only" in a sentence keeps changing the meanings. "Only" as an adjective modifying "truth" should be near this word. But "only" can also be an adverb. Strictly speaking, "I only told my supervisor the truth" is a squinting construction. It means either "Only *I* told my supervisor the truth" or "I only *told* (as opposed to writing or singing it) my supervisor the truth." That we do not know which of these improbable meanings is intended is not the point. Considering what we want to say, "only" was misplaced.

2. Over the weekend, he bought the chemical that was needed to complete his experiment. Or, if *this* meaning is intended: He bought the chemical that was needed to complete his experiment over the weekend.

Original: "He bought the chemical over the weekend that was needed to complete his experiment." This is one of those squinting modifiers since either of the two correct meanings can be intended. "Over the weekend" cannot be left "in the middle" like this.

3. The man lives with his son in a four-room house, which he rents for $150 a month.

Original: "The man lives in a four-room house with his son, which he rents for $150 a month." The adjectival "which . . ." clause was placed after the noun it modifies, "house," instead of after the noun that makes the reference funny, "son."

4. Because ether volatilizes, it should be stoppered when not in use.

Original: "Because ether volatilizes when not in use it should be stoppered." To permit the adverbial clause "when not in use" to modify the wrong verb, "volatilizes," suggests a false statement, namely, that ether does not volatilize merely when not used. But, more important, this clause position fails to express what is intended—the need for stoppering ether when not wanted. In its original position, the clause squinted toward the right meaning; but placing "when not in use" at the end removed all doubt.

5. The "expert" was asked to kindly resign.

Original: "The 'expert' was kindly asked to resign." Here "kindly" squinted toward both how he was asked and how it was hoped he would resign, namely, without making trouble. To express the latter meaning, which was no doubt the one intended, note that the only way to do it colloquially is to split the infinitive. "To kindly resign" is both expressive and proper; "to resign kindly" is formally correct but not acceptable as working English.

Answers to Exercise 15, p. 62

1. Newspaper headline: Stereo Set, Cash
                          Taken from Beauty
                          School by Burglars

Original: "Stereo Set, Cash
              Taken by Burglars
              From Beauty School"

In defending the original version, the writer would probably say that he knew his headline would sound funny but that it was the best he could do, since to write it as we have would have resulted in an 18-unit line, "School By Burglars"—one unit more than he was allowed. Does the idea of beautiful burglaries make sense? To correct, we keep "from Beauty School" from modifying "burglars" by placing it next to what it does modify, "cash taken."

2. I found, in the wrong folder, the report that praised him.

Original: "I found the report that praised him in the wrong folder." The prepositional phrase "in the wrong folder" should be placed after the verb it modifies, "found."

3. In machining there are two distinct operations, called reaming and broaching, done by experienced workmen.

Original: "In machining there are two distinct operations done by experienced workmen called reaming and broaching." To prevent its seeming to modify "experts," the expression "called reaming and broaching" should be placed after the noun it does modify, "operations."

4. You once told me she was a champion typist. *Or:* You told me she was once a champion typist.

Original: "You told me once she was a champion typist." Of course "once" squints toward either of the corrected meanings, one of which must be chosen.

5. The foreman promised to review the complaint as soon as possible.

Original: "The foreman promised as soon as possible to review the complaint." Clarity of meaning is secured by placing "as soon as possible" so it modifies "to review" instead of "promised."

6. I just had a split second to swerve the car to the right.

Original: "I had a split second to just swerve the car to the right." The adverb "just" modifies "had," so it should be placed near it instead of being allowed to split an inoffensive infinitive, "to swerve."

7. As we were leaving, he promised to order the printing.

Original: "He promised to order the printing as we were leaving." The chances are that "as we were leaving" refers to when he made his promise, rather than to the time at which he was to place the printing order.

8. He weighed on the scales the carefully prepared concentrate.

Original: "He weighed the concentrate on the scales that had been carefully prepared." It is a 100-to-1 bet that, although the scales had been carefully constructed at one time, this is not at all what is meant. Note that placing "on the scales" where it belongs permits us to drop three words, "that had been," to form a single independent clause instead of an independent and a dependent clause.

9. An empty process tube, a dummy fuel element, and two in-core instrument thimbles, which had been in the core during hot circulation tests, were successfully washed.

Original: "An empty process tube, a dummy fuel element, and two in-core instrument thimbles were successfully washed which had been in the core during hot circulation tests." Again, clarity is improved by placing the "which . . ." clause directly after the noun it modifies, "thimbles." The "which" clause also requires commas, since it is nonrestrictive in meaning. Note the fact that one need not understand the subject matter of a sentence to know that it is poorly constructed!

10. At least one suggestion prize announcement is made in almost every edition of our weekly newspaper.

Original: "At least one suggestion prize announcement is made in every edition of our weekly newspaper almost." The adverb "almost," modifying the adjective "every," should be placed just before it.

Answers to Exercise 16, p. 66

1. We gave each of them a chance to protest and then file a petition for transfer.

Original: "We gave each of them a chance to protest, and then they could file a petition for transfer." Since "to protest" and "file" are similar actions in this framework, they should be given parallel expression in the sentence. Why confuse the reader by presenting one action as an infinitive and the other as an independent clause?

2. He is one of those sensationalists who have given research a bad name.

Original: "He is one of those sensationalists who has given research a bad name." One of the five rules discussed on p. 64 was that verbs must agree with their subjects. So the verb "has" must agree with its subject, which is "who." But "who" is a pronoun which, being "pro-" or standing for a noun, must have a noun as referent. What is that noun? It could be either "sensationalists" (plural) or "one" (singular). Which is it? If it were "sensationalists," the correct verb form would be "have." If it were "one" the correct verb form would be "has." To see that "who" refers to "sensationalists," and therefore that the correct verb form is "have," we transpose mentally the entire modifying part of the sentence "he is one." This gives us, "Of those sensationalists who have (you see that, doing it this way, we would never say "has") given research a bad name, he is one."

3. First consider the origin of this theory and then its development.

Original: "First consider the origin of this theory and then how it has developed." This case resembles that of 1. "Origin" and "development" are similar things (you might review p. 63 on similarity), so they should be expressed similarly in the sentence. Thereby the reader can be confirmed in the idea that they *are* similar. Since "theory" is a noun, it should be followed by "development," another noun, instead of by a new sentence element, "how it has developed."

4. Mixtures of potash and sulphur require a heavy blow to explode them, whereas compounds of fulminate of mercury explode at a hard look.

Original: "Mixtures of potash and sulphur require a heavy blow to explode them, whereas you can explode compounds of fulminate of mercury by giving them a hard look." Since we are talking about two cases that are similar in that they refer to explosions, we should not discuss them in different persons, but use a parallel construction. ". . . you can explode" is in the second person. The entire sentence should be in the third person.

5. He proceeded methodically and took the necessary steps in proper order.

Original: "He proceeded methodically, and the necessary steps were taken in proper order." Can you feel the wrongness of this sentence? The sameness of action represented in the corrected sentence by "he proceeded" and "(he) took" is impossible in the original sentence because the second independent clause is in the passive voice. Instead of saying he took the steps, one says the steps were taken. The original sentence missed the chance to have the same subject ("he") for both clauses, and thereby did not express the parallelism of "his" actions.

Answers to Exercise 17, p. 66

1. I deal with the discount houses because of their lower prices and larger stock. *Or:* I deal with the discount houses because their prices are lower and their stock is larger.

Original: "I deal with the discount houses because of their lower prices and their stock is larger." The failure to have a grammatical parallelism reflect the duality of reasons for dealing with discount houses is repaired by making both reasons either prepositional phrases ("because of their lower prices") or dependent clauses ("because their prices are lower").

2. The Forbes Company plans to extend its holdings.

Original: "The Forbes Company plans to extend their holdings." Since pronouns must agree with their antecedents in number, "their" should be "its" because "the Forbes Company" is being considered as a single unit.

3. Darwin was admired by fellow scientists, but was hated and slandered by the clergy. *Or:* Fellow scientists admired Darwin, but the clergy hated and slandered him.

Original: "Darwin was admired by fellow scientists, but the clergy hated and slandered him." The duality of the attitudes toward Darwin should be confirmed by a duality in any sentence expressing these attitudes. The "was admired" sentence includes a "was hated" element—two passive voice expressions. In the original, "was admired" is in the passive, and "the clergy hated" is in the active voice. Of the two corrected versions, the first is preferable because it is more concise. The single subject "Darwin" is adequate for the compound predicate.

4. The laser is one of the most interesting inventions that have ever been made.

Original: "The laser is one of the most interesting inventions that has ever been made." "That" refers to "inventions" and thus requires a plural verb. Using our transposing device, we have, "one of the most interesting inventions that have ever been made, the laser is." Put in this way, we would never say "inventions that has . . . ."

5. Let us first review the cause of the trouble and its disposition. *Or:* Let us first review how the trouble was caused and how it was handled.

Original: "Let us first review the cause of the trouble and how it was handled." This sentence fails to make parallel expressions of two similar things related to "the trouble." The first amended sentence, however, uses two nouns, "cause" and "disposition;" the second amended sentence uses two "how" phrases.

6. Our plant has 20,000 workers; we are not Fascists, but we must provide moral education here, or we will turn out Communists.

Original: "Our plant has 20,000 workers; we are not Fascists, but one must provide moral education here, or you will turn out Communists." This sentence changes from the first person ("our plant" and "we") to the third person ("one"), then the second ("you"); it also changes from the plural ("our" plant and "we") to the singular ("one"). The amended sentence is consistent in person and number.

7. The stipulated task of cutting, drilling, and polishing was performed rapidly.

Original: "The stipulated task of cutting, drilling, and polishing were performed rapidly." The subject of the verb is the singular noun "task"; therefore the verb must be singular, "was" instead of "were." In general, the proximity of a noun or nouns to a verb should not be allowed to influence its number, which is dictated by that verb's *subject* only.

8. If anyone objects, I shall discharge him.

Original: "If anyone objects, I shall discharge them." This is colloquial, but incorrect. A pronoun should agree with its referent in gender, number, and person. "Anyone" is in the third person singular and is the referent for the pronoun. The second pronoun used, "him," should agree.

9. Businessmen face knotty problems in estimating costs and particularly in gauging demand.

Original: "Businessmen face knotty problems in estimating costs and particularly gauging demand." The parallelism in the two types of problems was impaired by leaving out the second "in." Though the careful writer often leaves out words to enhance an effect, he often carefully includes them for the same reason. One can develop an "ear" or "sentence sense" for these inclusions and exclusions.

10. Gallium can be melted like this, but not steel.

Original: "Gallium can be melted like this, but you cannot melt steel that way." The improved sentence saves four words, which is an extra dividend from staying consistently in the passive voice and in the third person. "Can be melted" is passive, whereas "you cannot melt" is active. "Gallium can be" is third person, whereas "you cannot melt" is second person. Voice and person have been made consistent, thus effecting a parallelism of treatment between the "Gallium" and the "steel" phrases. It is hard to appreciate this parallelism, the sentence being so short as well as elliptical.

THE 1400 "AURA" WORDS

A study of the following 1400 words will greatly improve the student's vocabulary—if this study is systematic and thorough. Since this is the most important of many exercises that could be done, we urge that one column a day be read, with the student marking and then looking up in the dictionary all words whose meanings are unknown to him. When all the words have been examined, the marked ones are to be restudied. Then the student should repeat the entire process in another month or so.

The student should pause over each word, take in its flavor, and allow it to sink in, without worrying about remembering it.

We call these "aura" words. An aura is "a distinctive atmosphere surrounding a given source." The atmosphere is the life of technical writing; the source is the technical writer.

abate	achieve	affix	amplitude
abbreviate	acknowledge	aforesaid	anachronism
aberration	acquiesce	afterward	analogous
abeyance		agency	analysis
ability	acquire	agglomeration	animate
abortive	acquisition	aggregate	annex
abridged	activity		annotate
abrogated	actuate	agree	annular
abrupt	acute	alienate	anomalous
absolute	adamant	allay	
	adapt	allege	anonymous
absorb	addendum	alleviate	antecedent
abstraction	adduce	allocate	antedate
abstruse	adept	allot	anterior
academic	adequate	allow	anticipate
accede	adjacent	alloy	antidote
accentuate	adjoin	allusion	antithesis
accessible	adjudge		aperture
accessory	adjunct	alter	apogee
accommodate	adjust	alternate	apparatus
accomplish	admissible	alternative	
	admixture	although	apparent
accordingly	advantage	ambient	appearance
account	advent	ambiguous	appellation
accredit		amenable	append
accretion	adventitious	amend	appendix
accrue	affect	amorphous	appertain
accumulate	affinity	amount	appliance
accurate	affirm	ample	applicable

apportion
apposite
appraise
apprehend
apprise
appropriate
approximate
approve
appurtenance
a priori
apropos
aptitude

arbiter
arbitrary
archaic
archetype
arduous
argument
arrange
arrogate
articulate
artifice

ascendant
ascribe
aspect
aspire
assay
assemble
assent
assertion
assess
assiduous

assign
assimilate
assist
associate
assumption
assured
astute
asunder
attach
attain

attention
attenuate
attitude
attract
attribute
attrition
atypical
augment
aura
auspicious
authentic
authoritative
authorize
automatic
autonomous
auxiliary
avail
average
averse
avid

avocation
avoid
axiom
balance
barrier
because
behavior
belated
belief
beneficial

besides
bias
biennial
binary
boundary
brevity
calculate
cancel
capability
capacity

cardinal
category
catholic

cause
certainty
certify
cessation
character
check
chief
circuitous
circulate
circumscribe
circumspect
circumstance
classify
coalesce
coexisting
cogent
cognizant

coherent
coincidental
collate
collective
collocate
colloquial
combine
commendable
commensurate
comminute

commit
commodious
common
communicate
commute
compare
compass
compatible
compeer
compel

compendium
compensate
competence
complement
complexity
compliance

complicated
component
compose
composite
composition
compound
comprehensive
compress
comprise
compute
concatenate
concede
concentrate
conception

conclusive
concrete
concur
condensation
condign
condition
condone
conducive
conduct
configuration

confirm
conform
congruent
conjunction
connect
connective
connote
conscientious
consecutive
consequence

consideration
consistent
consolidate
constitute
construct
construe
consummate
contaminate
contend

content
conterminous
contiguous
contingent
continual
continuity
continuous
contract
contradictory
contrary
contrast

convention
convergence
converse
convert
convey
convolution
coordinate
corollary
correct
correlate

correspond
corroborate
cosmos
counteract
counterpart
create
credible
credit
criterion
critical

crucial
cryptic
cultivate
cursory
curtail
data
dearth
decadent
decal
declare
decrease

decrement
deduction
de facto
defect
defer
deficiency
define
definite
definitive

deflect
defray
degree
de jure
delete
delineate
demarcate
demonstrate
denominate
denote

dependent
deplete
deprive
derive
derogatory
description
desideratum
design
designate
desultory

detach
detail
detect
deteriorate
determine
detract
development
deviation
device
devise
devoid
diagnose
dichotomy
dictate

diction
differentiate
difficult
diffuse
digress
dilatory

dilemma
dimension
diminish
direction
disadvantage
disarrange
discern
discharge
disclaim
discontinuous

discount
discovery
discrepancy
discretion
discriminate
discursive
disintegrate
disinterested
disjoin
disjunctive

disorder
disparity
dispersion
displace
display
dispose
disseminate
dissent
dissimilar
dissolve
dissuade
distinct
distinguish
distort
distribute
disunite
disuse

diverge
divers
diverse

diversified
divide
divulge
doctrine
document
dogma
dominant
drastic
dual
duplicate

durable
duration
dynamic
eclectic
economy
effective
efficient
effort
egregious
egress

elaborate
elapse
element
elicit
eliminate
elucidate
emanate
embellish
embody
emend
eminent
emit
emphasis
empirical
employ
empower
emulate
enable
enact
enclose

encumber
endeavor
enervate
engage
engineer
engross
enhance
enigmatic
enlist
ensue

entail
entertain
entire
entitle
entry
enumerate
environment
ephemeral
episode
equable

equity
equivalent
equivocal
eradicate
erratic
error
erudite
esoteric
essay
essence
essential
establish
estimate
ethical
etymology
evanescent
eventual
evidence
evince
evolution

evolve
exact
examine

example
exceeding
exception
excerpt
excessive
exchange
exclude

execute
exemplify
exemption
exercise
exert
exhaustive
exhibit
exigency
exiguous
existing

exonerate
exorbitant
expand
expedient
expedite
expend
experiment
explain
explicit
exposition
expound
express
expression
expunge
expurgate
extant
extempore
extenuate
extensive
exterior

external
extract
extraneous
extraordinary
extrapolate

extreme
extrinsic
fabrication
facet
facility

facsimile
fact
factitious
factor
faculty
fallacious
fallible
familiar
feasible
feature

figure
finite
fiscal
flagrant
flaw
flourish
fluctuate
fluent
follow
foregoing
foresee
forestall
forethought
form
formation
formula
fortuitous
fractional
fragmentary
frame

frequent
frustration
fulfill
function
furnish
furthermore
futile

gain
gamut
gauge

general
generate
genuine
genus
germane
gist
govern
gradation
gradual
graphic

gross
groundless
group
habitual
harmony
hazard
hence
herewith
heterogeneous
hiatus
hindrance
homogeneous
however
humane
hypothetical
idea
ideal
identical
identify
ideology

illuminate
illustrate
imagination
imitate
immanent
immediate
imminent
immutable
impalpable
impartial

impasse
impediment
impel
impending
imperative
imperceptible
impervious
impetus
impinge
implement

implication
implicit
import
impose
impress
improvement
improvise
impunity
impute
inadvertent
inalienable
inane
inasmuch
incalculable
incapacitate
inception
incessant
inchoate
incidental
incipient

incline
inclusive
incompetent
incongruous
incontrovertible
increase
increment
incumbent
indefinite
independent

index
indicate
indicative

indigenous
individual
induce
induction
indued
inept
inert

inextricable
infer
inferior
infinitesimal
inflexible
influence
inform
information
ingenious
ingenuous
ingredient
inherent
inhibit
inimical
inimitable
initial
innate
innocuous
innovation
inquiry

insert
inspect
instance
instantaneous
instruction
instrumental
insure
intact
integrity
intend

intensive
intent
interchange
interior
interpolate

interpose
interpretation
interval
intervene
intransigent

intricate
intrinsic
inure
invalidate
inventory
inversion
investigation
inveterate
invidious
involve
irregular
irrelevant
irrespective
isolate
issue
jargon
jeopardy
jointly
judgment
judicious

juncture
jurisdiction
justify
juxtaposition
kilo
kin-
knowledge
laconic
laissez faire
language

latent
lateral
latitude
laudatory
law
learning
legislate

length
level
liable

liberal
likely
likewise
limited
lineal
linear
literal
locate
locus
logic
logo-
longitude
lucent
lucid
luminous
machine
macro-
magnify
magnitude
maintain

major
management
mandatory
maneuver
manifestation
manifold
manner
manual
manufacture
marshal

mask
mass
master
material
matter
maturity
maximum
mean
means
measure

merit
metamorphosis
meter
method
micro-
mil-
minify
minimum
minor
minute
misnomer
miscellany
mitigate
mobility
mode
model
moderate
modicum
modify
modulate

momentary
momentous
monitor
mono-
monograph
moot
moreover
moribund
motif
motive

motley
mount
multi-
mundane
mutual
nebulous
necessary
negative
neglect
negotiate

neo-
neutral
nevertheless

node
nominal
nondescript
non sequitur
normal
notable
notion
notwithstanding
novel
nucleus
nullify
numerable
numerous
objective
oblige
oblique
obscure

observe
obsolete
obtain
obtuse
occasion
occupy
occur
offer
offhand
omit

onerous
operate
opinion
opportune
opposite
option
oral
order
ordinal
organize

origin
orthodox
oscillate
ostensible
outlet

outline
outward
overlook
oversight
paginate
pale-
palpable
pan-
panorama
par
para-
paradox
parallel
paramount
paraphrase

parenthetical
partial
partake
participate
particle
particular
partition
patent
pattern
paucity

pending
penetrate
per
perceive
percent
perceptible
peremptory
perennial
perfect
perforce

perform
periodic
permanent
permeable
permission
permutation
perplexity

persist
perspective
perspicuous
persuade
pertain
pertinent
pervade
phase
phenomenon
phil-
pioneer
plastic
plausible

plentiful
plurality
pneumatic
poignant
polemic
politic
polity
pollute
poly-
ponder

portend
portion
position
positive
possess
possible
post-
posterior
postpone
postulate

potent
power
practicable
practical
practice
pragmatic
pre-
precarious
precaution
precedent

preceding
precept
precise
preclude
preconception
precursor
predicament
predicate
predict
predilection

preeminent
preempt
preferable
preliminary
premature
premise
prepare
preponderance
prerequisite
prerogative

presage
prescribe
presence
present
preserve
presume
prevalent
prevent
previous
primary

prime
principal
principle
priority
privacy
pro-
probability
problem
procedure
process
procure
prodigious
produce

profess
proficient
progress
project
prolific
prolong
prominent

promote
prompt
promulgate
prone
proof
propagate
propensity
proper
propinquity
propitious

proportion
proposal
propose
proposition
propound
prosecute
prospective
protagonist
prototype
protract

protruberance
protrude
provide
province
provision
proviso
provocation
proximate
pseudo-
publish
punctual
punctuate
purge
purpose
puzzle
pyro-

qualify
quality
quantity
question

quota
quote
radial
radical
radius
ramification
range
rank
rapid
rate

ratify
ratio
rational
raw
raze
reaction
reasoning
recalcitrant
recant
recast

recede
reception
recession
recipient
reciprocal
reckon
reclaim
recognize
recommend
reconcile
reconstruct
record
recourse
recovery
rectify
recur
redintegrate
reduce

redundant
refer

refine
reflect
refractory
refute
regard
regenerate
regimen
register
regression
regular

regulate
rehabilitate
reinforce
reiterate
relationship
release
relegate
relevant
relinquish
reluctant

remand
remark
remedy
remind
remiss
remit
remunerate
render
renovate
repair
repel
repercussion
repetition
replace
report
represent
repress
reproduce
request
requisite

resemble
reserve
residual
resistant
resolve
resonant
resource
respectively
respite
response

restore
restrict
result
retain
retard
retract
retrograde
reversion
revert
review

revise
revolution
revolve
right
rigid
rigorous
rudiment
sagacious
salient
salutary
sample
sanction
satisfy
saving
scan
scant
scarce
scarcely
scarcity
scattered

schedule
schematic

science
scope
scrupulous
scrutiny
seclude
secrete
sector
security

sedentary
segment
segregate
select
semblance
seniority
sensation
sensible
sensitive
sensuous

sentient
separate
sequence
settle
several
severe
shade
shape
sharp
sign
significant
similar
simulate
simultaneous
since
situation
skeptical
skillful
slight
solicit

soluble
solution
solvent
sophisticated

source
specialize
specialty
species
specific
specification

specify
specimen
specious
spectrum
speculative
sphere
spontaneous
sporadic
spurious
stable

standard
step
stereo-
stimulate
stint
stipulate
stratum
strenuous
stress
striated
stringent
structure
subjective
submit
subordinate
subscribe
subsidiary
subsidy
subsist
substantiate

substitute
subsume
subversive
succeed
succinct
sufficient

suggest
suitable
summarize
sumptuary

sundry
super-
supercede
superficial
superfluous
superimpose
superior
superstructure
supplant
supplementary

supply
support
supposition
surmise
surplus
survey
suspend
sustain
symmetrical
synchronize
synonymous
synopsis
synthesis
system
tacit
tactile
tangent
tangible
tantamount
technique

tectonic
tele-
temperate
temporal
temporize
tenable
tenet
tenor

tentative	transform	universal	verbose
tenuous	transient	unqualified	verge
tenure	transition	untenable	verification
terminal	transmit	urgent	verisimilitude
terminate	transmute	usually	
terminus	transpire	utilitarian	vernacular
terrestrial		utility	versatile
terse	transport	utmost	version
test	transpose	vacant	vertical
text	traverse	vacillating	vestige
theory	treatise		vicarious
therefore	trenchant	vacuum	view
thesis	trivial	valid	vision
thorough	true	value	vivid
though	turgid	valve	volatile
thus	type	vantage	
titular	ultra-	variable	
tool	unanimous	variant	voluntary
topic	undefined	variation	vouch
total	undertake	variety	warrant
trace	unequivocal	various	whereas
traditional	undue	vary	whereby
	uni-	vent	withdraw
traffic	unified	venture	withhold
transact	unit	veracity	wrong
transcend	uniform	verbal	yield
transfer	unique	verbatim	zone

ANSWERS TO EXERCISES 18–21

Answers to Exercise 18, p. 69

1. This work is intended (or meant, planned, etc.) to support the main experiment.

Original: "This work is to support the main experiment." The sentence is ambiguous; it could mean merely, "This work is going to (or will) support the main experiment." But we feel that it means more than this and involves intention. If so, it should be said. In either case, "is" functions here as a verbal auxiliary, helping out the main verb. Not to supply the main verb makes "is" carry too much of the meaning load, besides being unclear.

2. The mixture was saturated with oil. *Or:* The mixture was nearly saturated with oil.

Original: "The mixture was more or less saturated with oil." What does "more or less" mean? If nothing, it should be deleted, and we get the first improved version. If it means "nearly," using this word would be a great improvement.

3. If personnel of your concern are convinced that the requirements should be revised, they may make recommendations through proper channels.

Original: "If the activities at your locality feel strongly that the requirements should be revised, they may recommend through proper channels their implications for the revision." Of course "activities" as such are not persons, and it is persons who are meant. There is no need to repeat the revision idea. "Implications" is misused; to recommend implications is meaningless.

4. Plans are to control the fuel cask operation remotely, viewing it through shielding windows.

Original: "The control of the fuel cask operation would be remote, and would be viewed through shielding windows." The sentence changes here are triggered by trying to make "would be" precise. Also, the original says the "control" is what is viewed, whereas, in the corrected version, "it" refers properly to the operation. Why did the final sentence not achieve a parallelism, to wit: "Plans are to control the fuel cask operation remotely, and to view it through shielding windows"? The answer is that this makes the controlling and the viewing coordinate in importance, whereas it was felt that the viewing should be subordinate.

5. A writer knows his readers will have differing degrees of familiarity with the report subject.

Original: "In writing any report, the readers have a different familiarity with the report subject." We must first decide whether the subject is "writer" or "readers." We felt it was "a writer" and that the subject dealt with his knowledge of his readers. What knowledge was meant? The original seemed ambiguous because its writer did not have the word "degrees" in his working vocabulary. The thought seemed, then, to be about degrees of knowledge.

Answers to Exercise 19, p. 70

1. This calculation has been made as part of another project.

Original: "This computation has been performed under another project." To the literal minded, "under" sounds odd, and technical persons should be literal minded. Also, "calculation" is a more precise term than "computation." Last, to "perform" a computation sounds pompous; it sounds almost like something done on a stage.

2. This experience has helped us greatly in writing equipment specifications.

Original: "The enlightenment of this experience has provided a firm basis on which to write equipment specifications." Remember that the presence of polysyllabic nouns at the beginning of a sentence may conceal the need for a strong verb. Which is the subject, "enlightenment" or "experience"? Is it not the latter? Can we not express the idea of enlightenment in the verb, meanwhile getting rid of the roundabout "provided a firm basis on which"? Fortunately we can. The strong verb that we need preexists in the language; it is "to help."

3. When the metal was rolled out to the required length, it became much too brittle.

Original: "While the metal was rolled out to the desired length, it became much too brittle." The writer did not want to call attention to the obvious fact that brittleness occurred during the rolling, but merely to state this result. So "when" is preferable to "while." In addition, "required" is a more objective and scientific term than "desired."

4. The can was designed to hold rods.

Original: "The can was designed for the use of rods." How can rods use a can? But the can is usable by humans, for holding the rods. This should be stated.

5. Checkout is not yet complete, but first results promise to agree more closely with the experiment.

Original: "Checkout is not yet complete, but first results are in a direction to provide greater agreement with experiment." The single word "promise" expresses what "are in a direction" is meant to say. And "to agree more closely" is more precise than "to provide greater agreement." Notice that here again an abstract noun, "agreement," has been rejected in favor of its verb form, "to agree."

6. A man's work should be judged by his equals.

Original: "A man's work should be referred to his peers as a standard." This is one of those "find the meaning" sentences, stimulating us to gentle detective work. We must reduce the original utterance to a significance one can live with. "Peers" must go; it is equivocal, meaning "superiors" in England and "equals" in America. "Referred to" is not even equivocal; it is ambiguous. We think, however, that the *words chosen* indicate this meaning: "A man's work should be accepted by his equals as a standard." This is nonsense, since one works to conform to a standard, the work and the standard being two different things. That anyone's work is automatically its own standard is ridiculous. So, as so often in rewriting, the editor tries to smoke out the writer's meaning. The question becomes, not "What do the words mean, so that I can say it better?", but "What in the world did the writer mean when he wrote this?"

7. In writing this proposal, we (or they, etc.) discovered significant factors and drew several conclusions.

Original: "The preparation of the above referred proposal pointed out several significant factors regarding the concept and resulted in several conclusions being drawn." A "preparation" cannot by itself point out anything. We had better choose a pronoun subject, "we" or "they," and give the gist of the two happenings involving "factors" and "conclusions."

8. The present design effort was intended to achieve a more gradual transition.

Original: "The present design effort was concentrated on providing a more gentle transition." If the subject is a mechanical one, "gentle" is out of place. "Was intended to achieve" is more precise than "was concentrated on providing," though it is not easy to say why. The mere saving of two syllables does not give the answer. Perhaps the reason is that "to achieve" sounds more like accomplishing something than does "on providing."

9. This was done in order to determine and control the conditions for testing carbides.

Original: "This was done in order to upgrade the conditions for testing carbides." The professional slang called jargon is universally used by workers in the arts and sciences. "Upgrade" is such a word, meaning generally to improve, or to raise to a higher classification. But jargon should not be used when one's readers are not colleagues; nor, certainly, when it leaves one's own colleagues guessing. This is the case here. A colleague would be apt to ask, "Just what do you mean?"

10. A new age dawned when man finally invaded space.

Original: "A new age dawned when they finally invaded space." This is an easy one to correct. "They" is imprecise.

Answers to Exercise 20, p. 74

1. This project comprises a fuel element design and a safeguards report.

Original: "The objective of this project is the design of a fuel element also to prepare a safeguards report." Since it is the nature of a project to have an objective, we need not say it. So we invoke Rule 1. To use "comprises" for our verb instead of "is" employs a more precise term (Rule 6). In the original, the two similar objectives were not similarly stated; this corrective involved Rule 5.

2. This poor writing—not entirely due to the press of schedules—should be improved.

Original: "Perhaps this poor writing is due to the press of schedules, but there surely should be some improvement along these lines." The writer had failed to state his thought: if the poor writing was due to lack of time, improvement could hardly be expected. He meant the opposite, that it was not *entirely* due to the lack of time. Marking this fact off with dashes features it as a poor excuse, and emphasizes the main point, that of needed improvement.

3. Adjusting the variable orifices presents two problems.

Original: "There appears to be two problem areas in regard to adjusting the variable orifices." The number of the verb should be similar to that of the subject noun, "areas"; had we not rearranged the sentence, we would have had to say "there appear." We dropped "in regard to," "appears," "areas" (Rule 1).

4. Above 1000°C, Hi-Tec pyrolytically decomposes.

Original: "Hi-Tec is limited to a maximum temperature of 1000°C due to pyrolytic decomposition at high temperatures." Rule 1 drastically applied: we assume that members of the target public for this sentence realize that to decompose constitutes a limit on usage. Nor need we repeat the high temperature idea.

5. His most biting comment is that young men who are helping to fight a war should not be subjected to such a humiliating experience.

Original: "His most biting comment is that young men who are helping fight a war should be subjected to such a humiliating experience." "Not" had been left out, and "helping to fight" is clearer than "helping fight." With these two additions, we avoided an incomplete construction (Rule 2).

Answers to Exercise 21, p. 74

1. The discrepancy between theory and experiment noted in the last quarterly report has been carefully examined.

Original: "The discrepancy between theory and experiment noted in the last previous quarterly report has been carefully examined." We do not need both "last" and "previous" (Rule 1).

2. The effect of high gas solubility on heat transfer is unknown.

Original: "The problem of high gas solubility and its effect on heat transfer is unknown." This sentence has a compound subject, "problem" and "effect," and should be followed by "are" rather than "is" (Rule 5). The first subject is presumed to be self-explanatory. But it is not. The writer admitted that he meant the corrected version. Emphasis (Rule 3) and precision (Rule 6) have been secured.

3. The detection of Fe, Cr, and Ni in thin film deposits and of slight surface irregularities in the hot-leg indicates (or suggests) that corrosion may have occurred.

Original: "The detection of Fe, Cr, and Ni in thin film deposits and of slight surface irregularities in the hot-leg imply corrosion may have occurred." The verb, having a singular subject, "detection," should be "implies" (Rule 5). This, however, is the wrong verb, since nothing implies that something *may* have occurred; implication, by definition, involves necessity. So "indicates" or "suggests" is preferable, depending on how strongly one wants to state the thought (Rule 6). Rule 5 is also invoked to parallel "*in* thin . . ." with "*in* slight. . . ."

4. The reason he failed was that he could not take criticism without losing his temper.

Original: "The reason he failed was he could not take criticism without losing his temper." Rule 2 was used. The added "that" aids the reader by indicating that the long clause, "he could not take criticism without losing his temper" is to be understood as a unit.

5. The alloyed ingots were analyzed (Table 4) for carbon and nitrogen and the alloying elements.

Original: "Chemical analyses for the alloyed ingots appear in Table 4. The alloyed ingots were analyzed for carbon and nitrogen and the alloying elements." Surely we do not need the first sentence; "(Table 4)" will serve. That the analyses were chemical would be obvious (Rule 1).

6. The study indicates that material erosion and corrosion data can be extrapolated to larger systems.

Original: "The study infers that material erosion and corrosion data can be extrapolated to larger systems." Only persons can infer. The study "indicates." Rule 6 has been employed.

7. The new employees will be here tomorrow and will be processed the day after.

Original: "The new employees will be here tomorrow, and processed the day after." It may be dull to repeat "will be," as the hasty, careless writer so decides; but the careful writer wants, above everything else, to be accurate and clear. He adds the second "will be" (Rule 2).

8. Because components were operated only for testing purposes, minimum maintenance was required.

Original: "Components have been operated only for testing purposes, resulting in low maintenance requirements." The cause-and-effect relationship can be brought out by converting the loose sentence to a suspended one (Rule 3). "Minimum" is probably more accurate than "low" (Rule 6). Also, "resulting in . . . requirements" has been tightened to "was required" (Rule 1).

9. We (or they, etc.) finally solved the problem.

Original: "Having at last found the solution, the problem was solved." Since the inanimate problem itself could certainly not find a solution, we have here a dangling modifier. This could be corrected by saying, "Having at last found the solution, we solved the problem (Rule 2). But the sentence has needless words; the solution need not be referred to twice (Rule 1).

10. These experiments on UC single crystals are intended to make some so close to flawlessness that X-ray microscopy can be applied.

Original: "The purpose of these experiments on improving the perfection of UC crystals is to prepare crystals that are sufficiently flawless to permit the use of X-ray microscopy." In evaluating a sentence, is there a first thing that one looks for? Perhaps it is whether words, phrases, or ideas seem to be repeated. We appear to have several such repetitions here. "Improving the perfection" and "sufficiently flawless," besides being self-contradictory, seem to mean the same thing (Rule 1).

The following sentences will be corrected without comment so that the reader can analyze the improvements by himself.

11. As a result of the engineering mock-up work, solutions to the following problems have been clearly indicated.

Original: "As a direct result of the engineering mock-up work, the following are problems to which feasible solutions are apparent."

12. As the hot rolling schedule indicates (Table 7), rolling temperatures varied with alloy content.

Original: "The hot rolling schedule is presented in Table 7. As the rolling schedule indicates, the rolling temperatures varied with the alloy content."

13. This procedure revealed that defects could be removed.

Original: "By this procedure it was established that defects could be removed."

14. A sealed chamber would greatly increase the pressure.

Original: "If the chamber is sealed, it will greatly increase the pressure."

15. I have always tried and will always keep trying to meet my deadlines.

Original: "I always have and will always keep trying to meet my deadlines."

SKILLFUL USE OF TRANSITIONAL PHRASES

The following address was made before a group of San Fernando Valley engineers by Ralph Balent, upon receiving the Engineer of the Year Award for 1964 "for outstanding contributions to the engineering profession and his community." Mr. Balent was Vice President, Power Systems Programs at Atomics International, in Canoga Park, California. This talk shows the skillful use of connectives. See how the italicized words help to integrate and clarify.

Atomic Energy—A Fertile Field for Creative Engineers

In 1905 Albert Einstein released his "Special Theory of Relativity" that stated his now universally accepted $E = mc^2$; i.e., energy and mass are equivalent. Nothing spectacular happened for 34 years—until early 1939. *In January 1939* two French theoretical physicists, in studying some odd results of an experiment utilizing uranium, proposed that the energy was due to the fissioning of uranium. *This news* rapidly spread throughout the international level of theoretical and experimental physicists, and Enrico Fermi (who had *just recently* left Fascist Italy and was at Columbia University) suggested the possibility of a nuclear chain reaction.

In late 1939, the services of Albert Einstein and Eugene Wigner were used to confirm the military potential of "fissioning" to President Roosevelt. *This meeting* gave birth to the now famous "Manhattan Project," which was formed to secretly develop the A-bomb.

In December 1942, at the University of Chicago, the world's first chain reaction was achieved.

Then in the summer of 1945, three A-bombs were detonated: the first at Alamogordo, the second at Hiroshima, and the third at Nagasaki.

After World War II, the interest in atomic power expanded to investigate other possibilities for nuclear power—*early emphasis* was on harnessing atomic power for military propulsion systems. The Navy initiated studies for nuclear submarines, and *this program* resulted in the successful launching of the Nautilus in 1954. This program has expanded ever since, and the Navy *now* has a sizeable fraction of its submarines nuclear-propelled.

Concurrently with Navy action in 1946, the Air Force let studies on nuclear propulsion for aircraft, ramjets, and rockets. *The first* grew into the so-called Aircraft Nuclear Propulsion, that is, the ANP program, which expended over $1 billion before it was cancelled by Eisenhower in 1960.

The nuclear ramjets and rockets were *first studied* by North American Aviation Corp. in 1946–48. *The results of these studies* indicated technical feasibility only upon successful high temperature materials development by the Air Force and were *later* picked up by the AEC in 1949. *The two projects* in the ground development phase bore the code names PLUTO and ROVER.

In the early 1950s, nuclear power was entering a third phase (first, bombs;

second, propulsion; third, power); *that is,* using nuclear heat for the production of electrical power.

Initially, power plus plutonium production was extensively studied. *Then,* when the gaseous diffusion plants and the production reactors of Hanford and Savannah River started to produce sufficient weapons materials—and some say more than enough—President Johnson reduced production rates of weapons materials for the first time since World War II, attention being directed at power only.

In the late 1950s, a number of experimental and prototype nuclear central station power plants were constructed throughout the United States. *Currently,* a number of these plants are producing approximately 1000 Mwe of power, which is being distributed in the networks of the electrical power grids throughout the United States. *However,* nuclear power still represents only a small fraction of the total electrical power production in the United States today. Approximately 3000 Mwe of nuclear stations are under construction in the U.S.; *thus,* it appears that these developmental and prototype reactors have led us to the point where we are on the threshold of economic nuclear power. *As a consequence,* we have extended the energy resources of the world over one hundredfold *when compared* with the energy potential of conventional hydrocarbon fuel supplies alone. *In addition,* research work is being developed to harness the fusion process—that is the process used in the hydrogen bomb—to control it for electrical power production. *When this is accomplished,* the earth will be forever free of a basic energy shortage.

A *fourth phase* of nuclear energy development was initiated *in the mid-50s.* First it was bombs; *second,* propulsion; *third,* central station electrical power. *The fourth phase* was for special purposes such as radiation sterilization of food, industrial irradiation of tracers and gauges, and *also* for small special-purpose power plants. I would like to take just a few moments to talk about one special-purpose power plant with which I am most familiar—that is, the SNAP systems.

Atomics International, at the suggestion of the Rand Corporation, initiated *in 1953* studies of very small reactors for auxiliary electrical power for space application. *After two years of preliminary studies* the Air Force and the AEC decided to let contracts for the development of small nuclear auxiliary power systems. A bidders' conference was held *in 1955.* Atomics International was awarded a contract *in 1956,* and achieved the first critical experiment in *October 1957*—that is, the same month the Russians ushered in the Space Age with the launching of Sputnik I. *Since that time,* these efforts have increased until SNAP (Systems Nuclear Auxiliary Power) *now* encompasses systems which cover the power range from a few hundred watts into the hundreds of kilowatt range.

This fourth phase of atomic energy for special purposes can expand in the future in a multitude of ways. Creative Engineering for the Designs of Tomorrow (which is this year's theme for Engineers Week) will probably include the development of many unique applications of atomic power. *For example,* atomic power could be used to attack the two major problems which currently are acute in the Greater Los Angeles area—air pollution and the ever increasing demand for fresh water.

In the first case, nuclear power could eventually be used to result in a smog-free city. This could be done in two ways.

1. Nuclear power plants could be constructed to supply a network of battery replacement stations at the same location where we *presently* see gasoline filling stations. The automobiles of the future would naturally have to be converted. to electrical power drives with a yet-to-be-developed highly efficient second battery. *With this scheme,* a person would drive up to a battery-replacement station, and by means of perhaps a hydraulic mechanism, a 300- to 500-lb battery with full charge could be exchanged for a spent battery. The person could *then* drive for his usual 200-mile run before requiring a replacement. *During this time* the spent battery could be undergoing electrical recharging from power supplied to the re-placement station by central station nuclear power plants.

2. *A second possibility* of utilizing nuclear power to eliminate air pollu-tion, which should receive some positive consideration, is the use of low-cost heat energy from a large nuclear plant as a source of heat for the production of synthetic smog-free fuel. *This synthetic fuel* could be distributed in the conventional manner to filling stations and burned in a modified internal combustion engine.

With regard to fresh water, studies at Oak Ridge National Laboratory and other sites have recently shown that large dual-purpose electrical plant and salt-water-desalinization plants have the potential for producing economic electricity along with economical fresh water as a by-product.

The development of smog-free cities and an abundant supply of electricity and fresh water would *then* release the creative engineer and designer to expend more energy in the exploration of space. It is conceded by all experts in the field that nuclear power for propulsion and on-board electrical power supplies is the only conceivable way by which man will be able to go beyond the present APOLLO moon project and explore our neighbor-ing planet Mars. It is not inconceivable that he will do this before the end of the present century, which is just a little over 35 years from now and only 95 years since Einstein released his famous equation, $E = mc^2$. However, it will require an ever-increasing number of creative engineers to accomplish these tasks.

ANSWERS TO EXERCISES 22-23 AND SUGGESTED FINALS

Answers to Exercise 22, p. 97

1. The obvious comment is that there are two paragraphs, whereas there should be one, since the one-sentence pargraph should be the first, and topic, sentence of the second paragraph. Everything in the second paragraph is an exemplification of the topic sentence, so the principle of development used is that of example.

2. The topic sentence is evidently the last one, the principles of development being temporal order and details.

3. To chop this long paragraph up into the various percentage parts would make it still more difficult to grasp. It is correct the way it is and extremely interesting. The material is just hard to learn—not a fault of the writer.

The topic sentence is the first one, and the principles of development are analysis and details.

4. The topic sentence seems to be the last one. If the principle of development were proof, we would be able to detect an argument. There is no argument, merely a statement of a situation. Because we tend to agree with the statements made, it seems as though reasoning were being done. This is not strictly so; the author is analyzing.

5. This long paragraph might better have been divided as follows.

Authors of the highest eminence seem to be fully satisfied with the view that each species has been independently created. To my mind it accords better with what we know of the laws impressed on matter by the Creator, that the production and extinction of the past and present inhabitants of the world should have been due to secondary causes, like those determining the birth and death of the individual.

When I view all beings not as special creations, but as the lineal descendants of some few beings which lived long before the first bed of the Cambrian system was deposited, they seem to me to become ennobled.

Judging from the past, we may safely infer that not one living species will transmit its unaltered likeness to a distant futurity. And of the species now living very few will transmit progeny of any kind to a far distant futurity; for the manner in which all organic beings are grouped shows that the greater number of species in each genus, and all the species in many genera, have left no descendants, but have become utterly extinct. We can so far take a prophetic glance into futurity as to foretell that it will be the common and widely-spread species, belonging to the larger and dominant groups within each class, which will ultimately prevail and procreate new and dominant species.

As all the living forms of life are the lineal descendants of those which lived long before the Cambrian epoch, we may feel certain that the ordinary succession by generation has never once been broken, and that no cataclysm has desolated the whole world. Hence we may look forward with some confidence to a secure future of great length. And as natural selection works solely by and for the good of each being, all corporeal and mental endowments will tend to progress towards perfection.

Thus we would get four paragraphs from the original single one. For the first paragraph, the topic sentence would be the second one; for the second paragraph, the entire sentence; for the fourth paragraph, the second sentence. What would it be for the third paragraph?

As for the principles of development used, proof is alluded to in the first paragraph ("accords better"), and is more strongly indicated in the "we may safely infer" and "hence" of the third and fourth paragraphs. Also, analysis is being applied.

Answers to Suggested Final for Chapter III, p. 100

1. This paragraph should not be broken up for the sake of clarity. The topic sentence seems to be the second one. The principle of development is analysis.

2. This paragraph is well integrated, each sentence seeming to be a new expression of the previous one, making it easier to grasp the topic sentence—the first one. Is this not an expanded definition of what Wells means by "universal education"?

3. Originally, there were three paragraphs, the second one starting, "some years ago I wrote" and the third one starting, "the statement is equally valid." This tripartite division was much better than the lumped-together paragraph we devised. The frequent use of "intellectual" was hard to avoid, since there is no good synonym in usage. "Mental," "learned," "cerebral," and so forth, do not convey the sense. But a paragraph preceding the three might well have been devoted to defining "intellectual."

4. The topic sentence is the first one. One principle of development used is contrast, as between animals and man. Time order, example, and analysis are also blended into this paragraph. It is difficult to grasp because of its philosophical nature; that is, characteristic high levels of generalization are involved.

5. Here Smith seems to be ignoring or forgetting the *intrinsic value* of what is worked on; otherwise merely working on something, from mud-pies to diamonds, would create equal values if the working times were equal.

He could not have meant this, therefore the reasoning of the paragraph is faulty. The topic sentence is the last one.

6. The "true" or representative paragraph (excluding purely transitional and one-sentence paragraphs) is one with a topic sentence, at least one detectable principle of development, and enough transitional words to hold the sentences together. The paragraph in name only is merely indented; the constitutive sentences do not cohere.

7. Your brief essay should consist of the textual remarks dealing with the role played by the topic sentence in the paragraph. It can begin the paragraph or appear elsewhere. It can be repeated.

8. This list is presented on the left-hand side of p. 84.

9. This exercise is up to the student, as is 10.

Answers to Exercise 23, p. 135

1. A period is to be placed after an abbreviation when it otherwise spells a word, as "in." for inch, or "gal." for gallon.

2. To master punctuation tends to give confidence to a writer because the 10 points involved apply to the entire language.

3. Points, or punctuation marks, are symbols that are meanings that we add to words. Thus a period is a little mark that indicates the end of a sentence. A question mark indicates an interrogation, at the end of which a speaker's voice would no doubt rise in pitch. Each point has its own meaning or meanings.

4. Yes to the first question, no to the second.

5. It often does. For example, Jack Spratt could eat no fat; his wife, no lean.

6. One can show that certain phrases and clauses in sentences are not essential by using commas before and after them.

7. Independent clauses can be separated without using coordinating conjunctions by using semicolons. For example, "He lost his temper; I kept mine." Here the semicolon takes the place of "but."

8. Semicolons precede sentence connectors, as: "The switch was open; *therefore* we closed it."

9. The point used to precede a specifying phrase such as "for example" is the semicolon, with a comma following the phrase. "He acted properly; that is, he filed the report."

10. The colon introduces a series, an explanation, a long quotation, a contrasting clause, or the body of a letter.

11. To stress a phrase, a dash may be used before it.

12. The basic rule for dividing words is to do it "by syllables."

13. Words should almost always be divided after vowels, and between consonants.

14. An apostrophe indicates a grammatical contraction and a coined plural.

15. Colons and semicolons are placed outside a closing quotation mark. Yes, it is always correct to put a period or a comma inside a closing quotation mark.

Answers to Suggested "Final" for Chapter IV, p. 146

The things to be learned from Chapter IV are not so much specific skills as they are salient facts and basic routines concerned with getting out a report. The emphasis has therefore not been placed on drill, but on the student's reacquainting himself with the text. As usual, we indicate where answers to the questions may be found in the text.

1. This subject is discussed briefly on p. 104. We want the student to realize how widespread the reporting function is, and therefore how important it is that he be skilled in it.

2. Two reliable tests are given on p. 107.

3. The guide sheet (p. 108) specifically directs the writer as to *how* the supervisor wants the assigned subject to be treated. The idea of the guide sheet, as amended by individual company needs, is invaluable. It is important that the writer not be overawed by the appearance of one, with its bristling entries. He could even invent and compose one of his own after being settled a bit in his work.

4. It is valuable to see why the list of elements (p. 110) might legitimately vary from one type of report to another, as needs vary.

5. The most difficult part of a report to write is the abstract. Poor abstracts, and why they are poor, may well be studied (p. 111).

6. To answer this involves either imagination or research (p. 114).

7. This discussion begins on p. 116.

8. This checklist (p. 117) should become part of the student's report-writing equipment.

9. The distinction between active and passive voices is frequently a difficult one for students to understand without a thorough explanation from an instructor (p. 120).

10. It would be interesting to learn to what extent a student has heard about checklists before (p. 122).

11. Reread p. 124–125.

12. These three topics are format, terminology, and punctuation (p. 125). The last topic belongs also to the subject of literary style.

13. Prefixes are written "solid," that is, without hyphens (p. 130). For example, it is "deenergize," not "de-energize."

14. The photo-offset method of printing (pp. 136, 140) may be contrasted with the usual newspaper type of printing, namely letterpress.

15. It is important to stress the fact that any type of publishing procedure has many variations, depending upon differing needs and personnel. One great aid in working out the best flow chart or publishing procedure is to have weekly meetings at which notes can be compared and complaints aired by those concerned. (p. 136).

16. This question concerns the original manuscript ("rough draft"), not the typed "repro" pages. The former is "edited," the latter "proofed." (p. 139).

17. A few editors prefer to proofread all by themselves. Since the removal or minimization of error is of course the aim, every editor should adopt the method that is best *for him*. (p. 140).

18. The correction sheet, too, is capable of variation, depending upon its most effective form locally (p. 142).

19. Development and use of a printshop checklist tends to sophisticate one as to publication problems, as the importance of item after item is understood (p. 143).

20. A good editor is apt to have several valuable character traits. As a rule, he spells trouble only to careless writers, and improves the writing products with which he concerns himself (p. 144).

Answers to Suggested "Final" for Chapter V, p. 157

1. "The Immediate and Growing Need for Specs," especially the first two paragraphs, answers this one (p. 148).

2. The classifications are given in "Hardware Specs and Writing Specs" (p. 149), "General and Detailed Specs" (p. 151), and "Specs Classified by Source" (p. 152).

3. The thoughtfulness of your answer is what is important. There could of course be several good answers.

4. "On Writing Specs Themselves" (p. 155) must be contrasted with the rest of the chapter to answer this. Writing *to* spec is obeying the dictates of a spec, carefully following its stipulations. *Writing* a spec is drawing up such stipulations, organizing a set of directions for others to follow.
5. This question, like No. 3, is designed simply to start one thinking. There is no single answer. Any good answer must show a definite feeling for organization.

Index